# UNDERSTANDING NARRATIVES FOR NATIONAL SECURITY

## PROCEEDINGS OF A WORKSHOP

Elizabeth Townsend, *Rapporteur*

Board on Behavioral, Cognitive, and Sensory Sciences

Division of Behavioral and Social Sciences and Education

*The National Academies of*
SCIENCES · ENGINEERING · MEDICINE

THE NATIONAL ACADEMIES PRESS
*Washington, DC*
www.nap.edu

THE NATIONAL ACADEMIES PRESS  500 Fifth Street, NW  Washington, DC 20001

This activity was supported by Contract No. 10003166 between the National Academy of Sciences and the Office of the Director of National Intelligence. Any opinions, findings, conclusions, or recommendations expressed in this publication do not necessarily reflect the views of any organization or agency that provided support for the project.

International Standard Book Number-13:   978-0-309-47639-3
International Standard Book Number-10:   0-309-47639-9
Digital Object Identifier:   https://doi.org/10.17226/25119

Additional copies of this publication are available for sale from the National Academies Press, 500 Fifth Street, NW, Keck 360, Washington, DC 20001; (800) 624-6242 or (202) 334-3313; http://www.nap.edu.

Copyright 2018 by the National Academy of Sciences. All rights reserved.

Printed in the United States of America

Suggested citation: National Academies of Sciences, Engineering, and Medicine. (2018). *Understanding Narratives for National Security: Proceedings of a Workshop*. Washington, DC: The National Academies Press. doi: https://doi.org/10.17226/25119.

*The National Academies of*
## SCIENCES · ENGINEERING · MEDICINE

The **National Academy of Sciences** was established in 1863 by an Act of Congress, signed by President Lincoln, as a private, nongovernmental institution to advise the nation on issues related to science and technology. Members are elected by their peers for outstanding contributions to research. Dr. Marcia McNutt is president.

The **National Academy of Engineering** was established in 1964 under the charter of the National Academy of Sciences to bring the practices of engineering to advising the nation. Members are elected by their peers for extraordinary contributions to engineering. Dr. C. D. Mote, Jr., is president.

The **National Academy of Medicine** (formerly the Institute of Medicine) was established in 1970 under the charter of the National Academy of Sciences to advise the nation on medical and health issues. Members are elected by their peers for distinguished contributions to medicine and health. Dr. Victor J. Dzau is president.

The three Academies work together as the **National Academies of Sciences, Engineering, and Medicine** to provide independent, objective analysis and advice to the nation and conduct other activities to solve complex problems and inform public policy decisions. The National Academies also encourage education and research, recognize outstanding contributions to knowledge, and increase public understanding in matters of science, engineering, and medicine.

Learn more about the National Academies of Sciences, Engineering, and Medicine at www.nationalacademies.org.

*The National Academies of*
## SCIENCES · ENGINEERING · MEDICINE

**Consensus Study Reports** published by the National Academies of Sciences, Engineering, and Medicine document the evidence-based consensus on the study's statement of task by an authoring committee of experts. Reports typically include findings, conclusions, and recommendations based on information gathered by the committee and the committee's deliberations. Each report has been subjected to a rigorous and independent peer-review process and it represents the position of the National Academies on the statement of task.

**Proceedings** published by the National Academies of Sciences, Engineering, and Medicine chronicle the presentations and discussions at a workshop, symposium, or other event convened by the National Academies. The statements and opinions contained in proceedings are those of the participants and are not endorsed by other participants, the planning committee, or the National Academies.

For information about other products and activities of the National Academies, please visit www.nationalacademies.org/about/whatwedo.

**STEERING COMMITTEE ON UNDERSTANDING NARRATIVES FOR NATIONAL SECURITY PURPOSES: A WORKSHOP**

CARMEN MEDINA (*Chair*), MedinAnalytics, LLC
SARA COBB, School for Conflict Analysis and Resolution, George Mason University
BETTY SUE FLOWERS, Department of English, University of Texas, Austin
JEFFREY C. JOHNSON, Department of Anthropology, University of Florida
DAVID MATSUMOTO, Department of Psychology, College of Science and Engineering, San Francisco State University
DOUG RANDALL, Protagonist

SUJEETA BHATT, *Study Director*
ELIZABETH TOWNSEND, *Research Associate*
RENÉE L. WILSON GAINES, *Senior Program Assistant*

## COMMITTEE ON A DECADAL SURVEY OF SOCIAL AND BEHAVIORAL SCIENCES FOR APPLICATIONS TO NATIONAL SECURITY

PAUL R. SACKETT (*Chair*), Department of Psychology, University of Minnesota
GARY G. BERNTSON, Psychology Department, Ohio State University
KATHLEEN M. CARLEY, School of Computer Science, Institute for Software Research International, Carnegie Mellon University
NOSHIR S. CONTRACTOR, McCormick School of Engineering and Applied Science, School of Communications, and Kellogg School of Management, Northwestern University
NANCY J. COOKE, The Polytechnic School, Fulton Schools of Engineering, Arizona State University
BARBARA ANNE DOSHER, Department of Cognitive Science, University of California, Irvine
JEFFREY C. JOHNSON, Department of Anthropology, University of Florida
SALLIE KELLER, Biocomplexity Institute, Virginia Polytechnic Institute and State University, National Capital Region
DAVID MATSUMOTO, Department of Psychology, College of Science and Engineering, San Francisco State University
CARMEN MEDINA, MedinAnalytics, LLC
FRAN P. MOORE, CENTRA Technology, Inc.
JONATHAN D. MORENO, Department of Medical Ethics and Health Policy, Perelman School of Medicine, University of Pennsylvania
JOY ROHDE, Gerald R. Ford School of Public Policy, University of Michigan
JEFFREY W. TALIAFERRO, Department of Political Science, Tufts University
GREGORY F. TREVERTON, Dornsife College of Letters, Arts, and Sciences, School of International Relations, University of Southern California
JEREMY M. WOLFE, Brigham and Women's Hospital, Departments of Ophthalmology and Radiology, Harvard Medical School

SUJEETA BHATT, *Study Director*
ALEXANDRA BEATTY, *Senior Program Officer*
JULIE ANNE SCHUCK, *Program Officer*
ELIZABETH TOWNSEND, *Research Associate*
RENÉE L. WILSON GAINES, *Senior Program Assistant*

# BOARD ON BEHAVIORAL, COGNITIVE, AND SENSORY SCIENCES

SUSAN T. FISKE (*Chair*), Department of Psychology and Woodrow Wilson School of Public and International Affairs, Princeton University
JOHN BAUGH, Department of Arts & Sciences, Washington University in St. Louis
LAURA L. CARSTENSEN, Department of Psychology, Stanford University
JUDY DUBNO, Department of Otolaryngology-Head and Neck Surgery, Medical University of South Carolina
JENNIFER EBERHARDT, Department of Psychology, Stanford University
ROBERT L. GOLDSTONE, Department of Psychological and Brain Sciences, Indiana University
DANIEL R. ILGEN, Department of Psychology, Michigan State University
NANCY G. KANWISHER, Department of Brain and Cognitive Sciences, Massachusetts Institute of Technology
JANICE KIECOLT-GLASER, Department of Psychology, Ohio State University College of Medicine
BILL C. MAURER, School of Social Sciences, University of California, Irvine
STEVEN E. PETERSEN, Department of Neurology and Neurological Surgery, Washington University School of Medicine
DANA M. SMALL, Department of Psychiatry, Yale Medical School
TIMOTHY J. STRAUMAN, Department of Psychology and Neuroscience, Duke University
JEREMY M. WOLFE, Brigham and Women's Hospital, Departments of Ophthalmology and Radiology, Harvard Medical School

BARBARA A. WANCHISEN, *Director*
THELMA COX, *Program Coordinator*

# Acknowledgments

This Proceedings of a Workshop was reviewed in draft form by individuals chosen for their diverse perspectives and technical expertise. The purpose of this independent review is to provide candid and critical comments that will assist the National Academies of Sciences, Engineering, and Medicine in making each published proceedings as sound as possible and to ensure that it meets the institutional standards for quality, objectivity, evidence, and responsiveness to the charge. The review comments and draft manuscript remain confidential to protect the integrity of the process.

We thank the following individuals for their review of this proceedings: Michael Schrage, Sloan School of Management, Massachusetts Institute of Technology, and Leah C. Windsor, Department of Political Science, University of Memphis.

Although the reviewers listed above provided many constructive comments and suggestions, they were not asked to endorse the content of the proceedings, nor did they see the final draft before its release. The review of this proceedings was overseen by Philip E. Rubin, Haskins Laboratories, New Haven, Connecticut. He was responsible for making certain that an independent examination of this proceedings was carried out in accordance with standards of the National Academies and that all review comments were carefully considered. Responsibility for the final content rests entirely with the rapporteur and the National Academies.

Carmen Medina, *Chair*
Steering Committee on Understanding Narratives
for National Security Purposes: A Workshop

# Contents

**1 INTRODUCTION**     1
The Decadal Survey of Social and Behavioral Sciences for
    Applications to National Security, 1
Objectives for the Six Workshops, 3
Introduction to the Workshop on Understanding Narratives for
    National Security Purposes, 4
Structure of This Proceedings, 4

**2 NARRATIVE RESEARCH IN THE SOCIAL AND
BEHAVIORAL SCIENCES**     5
Identifying Narrative Structure with Quantitative Methods, 5
Advancing Narrative Research with Big Data Science, 7
Function Words and the Arc of Narrative, 8
The Small-Story Approach to Narrative Research, 9
Discussion, 10

**3 THE SCIENCE OF NARRATIVE COMMUNICATION**     13
Narrative Processing and Sensemaking, 13
Narrative, Culture, and Digital Technology, 15
Communicating and Sensemaking Through Compressed
    Narratives, 16
Discussion, 17

| | | |
|---|---|---|
| 4 | **NARRATIVE AND EMERGING TECHNOLOGIES**<br>Computational Modeling of Narratives, 21<br>Attitudes, Behavior, and Influence, 22<br>Discussion, 22 | 21 |
| 5 | **NARRATIVE AND POWER**<br>Competing Narratives, 27<br>Countering Toxic Narratives, 28<br>The Power of Narrative, 30<br>Discussion, 31 | 27 |
| 6 | **SUMMATIVE COMMENTS**<br>Remarks from Karen Monaghan, 35<br>Remarks from Josh Kerbel, 36<br>Discussion, 38 | 35 |

**APPENDIXES**

| | | |
|---|---|---|
| A | Statement of Task for the Decadal Survey of Social and Behavioral Sciences for Applications to National Security | 41 |
| B | Workshop Agenda | 43 |
| C | Participants List | 47 |
| D | Biographical Sketches of Steering Committee Members and Presenters | 53 |

# 1

# Introduction

The Office of the Director of National Intelligence (ODNI), which oversees and directs the work of the 17 agencies and organizations responsible for foreign, military, and domestic intelligence for the United States, has a growing interest in research from the social and behavioral sciences (SBS) that may be beneficial to the Intelligence Community (IC). To develop a systematic understanding of these potential benefits, ODNI requested that the National Academies of Sciences, Engineering, and Medicine conduct a decadal survey of SBS to identify research opportunities that show promise for supporting national security efforts in the next 10 years.

## THE DECADAL SURVEY OF SOCIAL AND BEHAVIORAL SCIENCES FOR APPLICATIONS TO NATIONAL SECURITY

A decadal survey is a method for engaging members of a research community to identify lines of research with the greatest potential utility in the pursuit of a particular goal. The National Academies pioneered this type of survey with a study of ground-based astronomy in 1964.[1] Since then, committees appointed by the National Academies have conducted more than 15 decadal surveys. The Decadal Survey of Social and Behavioral Sciences for Applications to National Security represents the first opportunity to apply this approach to SBS. Its purpose is to develop an understanding of the lines of research in these fields that offer the greatest potential to enhance

---

[1] National Academy of Sciences. (1964). *Ground-Based Astronomy: A Ten-Year Program.* Washington, DC: National Academy Press. doi: https://doi.org/10.17226/13212 [April 2018].

the capabilities of the IC. To carry out this work, the National Academies appointed the Committee on a Decadal Survey of Social and Behavioral Sciences for Applications to National Security (Decadal Survey Committee); the committee's charge appears in Appendix A.

The Decadal Survey Committee has pursued many avenues in collecting information about the needs of the IC and relevant cutting-edge SBS research. As part of its information-gathering process, the committee held a series of six workshops—the first three on October 11, 2017, and the second three on January 24, 2018.[2] These workshops, for which planning began early in the committee process, were designed to explore areas about which the committee wished to learn more and to allow the committee to engage with a broad range of experts. The topics selected for the workshops do not necessarily indicate the ultimate direction of the committee's deliberations. The six topics addressed by the workshops were

1. changing sociocultural dynamics and implications for national security;
2. emerging trends and methods in international security;
3. leveraging advances in social network thinking for national security;
4. learning from the science of cognition and perception for decision making;
5. workforce development and intelligence analysis; and
6. understanding narratives for national security purposes.

Separate steering committees, whose membership included both members of the Decadal Survey Committee and additional experts in the topics to be addressed, were appointed to plan these workshops. Each of these committees was guided by its own charge. All were asked to bring their expertise to bear in identifying specific areas of promising research and experts with deep knowledge who could offer a range of insights.

This Proceedings of a Workshop, prepared by the workshop rapporteur, summarizes the presentations and discussions at the sixth workshop, on understanding narratives for national security purposes.[3] This workshop was planned by the Steering Committee on Understanding Narratives for National Security Purposes, whose charge is presented in Box 1-1. The workshop's purpose was to explore the current state of research on understanding narrative in the national security context. It should be noted that the steering committee's role was limited to planning and convening the workshop, and that the views contained in this proceedings are those

---

[2] For more information about the Decadal Survey and all of the workshops, see http://nas.edu/SBSDecadalSurvey [March 2018].

[3] The archived webcast of the workshop and available presentations can be found at http://sites.nationalacademies.org/DBASSE/BBCSS/DBASSE_184655 [April 2018].

> **BOX 1-1**
> **Workshop Steering Committee Charge**
>
> An ad hoc steering committee will plan and conduct a 1-day public workshop. The workshop will feature invited presentations and discussions to examine the nature and role of narratives as causes, effects, and transformers of cultures and people. The committee will plan and organize the workshop, select speakers and discussants, and moderate the discussions at the workshop. The workshop will be part of a set of workshops designed to gather information for the Decadal Survey of Social and Behavioral Sciences for Applications to National Security. A Proceedings of the Workshop will be prepared by a designated rapporteur in accordance with institutional guidelines.

of individual workshop participants and do not necessarily represent the views of all workshop participants, the steering committee, or the National Academies. The agenda for the workshop appears in Appendix B; a list of individuals who attended the three workshops held on January 24, 2018, is presented in Appendix C; and biographical sketches of the steering committee members and speakers are provided in Appendix D.

## OBJECTIVES FOR THE SIX WORKSHOPS

In an opening session for the three January 24, 2018, workshops, the chair of the Decadal Survey Committee, Paul Sackett, University of Minnesota, and sponsor representative William "Bruno" Millonig, ODNI, provided background information on the objectives for the six workshops.

Sackett observed that the Decadal Survey Committee will rely heavily on input from experts in the communities of national security and behavioral and social science research. Given the breadth of the committee's charge, he explained, it must cast a wide net, extending well beyond the specific expertise of its members to seek feedback from many sources. He described the six workshops as an important part of the effort to gather ideas. The workshops would support the committee by helping to identify promising research areas and allowing the committee members to engage in discussion with experts in a wide range of areas salient to its work.[4]

Millonig expressed appreciation to all those contributing to the committee's work through the workshops and other activities, noting that the participation of the full range of experts in the intelligence and behavioral

---

[4] Other activities associated with the Decadal Survey include calls for white papers, public meetings, and an online discussion forum; see http://nas.edu/SBSDecadalSurvey [March 2018].

and social science communities would be needed to make the decadal study successful. His remarks focused on the importance of SBS to the development of artificial intelligence, machine learning, and other automated tools. As an example of the value of such research, he noted that research on modeling behaviors and interactions is "fundamental to our ability to move forward [in utilizing these tools]." The research discussed at the workshops, he said, will help the IC understand the current and future contributions of these sciences.

## INTRODUCTION TO THE WORKSHOP ON UNDERSTANDING NARRATIVES FOR NATIONAL SECURITY PURPOSES

The history of narrative and storytelling ranges from prehistoric cave paintings to modern-day social media hashtags, remarked Carmen Medina in opening the workshop. As an employee of the Central Intelligence Agency (CIA) for 32 years, Medina has a clear understanding of the connection between narrative and national security. For example, she noted, nonstate actors often use stories as a tool for attracting followers and discrediting enemies. She added that narrative has also had a significant impact on such international events as the United Kingdom's 2017 Brexit election.

Medina pointed out that the steering committee for the workshop designed it to encourage discussion among the panelists, who each have different approaches to understanding and working with narratives based on the business or academic disciplines they represent. She added that the workshop was also designed to encourage lively discussion between the panelists and the audience, noting that after each of the panels had completed its presentations, the audience would have 15 minutes to make comments and ask questions. Furthermore, she explained, the committee had invited members of the IC to participate as active listeners and offer their thoughts on the discussions held throughout the day. The event would end, she noted, with an hour-long discussion open to everyone in attendance.

## STRUCTURE OF THIS PROCEEDINGS

This proceedings follows the structure of the workshop. Chapter 2 summarizes the workshop presentations and discussions on narrative research in SBS. Chapter 3 turns to the science of narrative communication. Chapter 4 explores emerging technologies and how human–computer interactions influence the flow of narrative. Chapter 5 focuses on the relationship between narrative and power. Finally, Chapter 6 includes the reflections of a panel of career intelligence analysts on the presentations summarized in Chapters 2 through 5.

# 2

# Narrative Research in the Social and Behavioral Sciences

The first session of the workshop was designed to set the stage for the day by providing an overview of research related to narratives in the social sciences and humanities, explained moderator Jeffrey Johnson, University of Florida. He noted that the four panelists had been asked to discuss the state of the art in this research from the perspectives of diverse fields, and to explore ways in which new technologies, particularly big data (very large datasets that can be mined or analyzed using only computer technology), can expand the possibilities for analysis of narrative.

## IDENTIFYING NARRATIVE STRUCTURE WITH QUANTITATIVE METHODS

Although the use of quantitative methods in narrative research is common practice today, this was not always the case, according to Roberto Franzosi, Emory University. A pioneer in quantitative narrative analysis, Franzosi explained that he first became interested in narrative structure while using statistical models to study time series data on the labor strikes of post–World War I Italy. At the conclusion of his research, however, he concluded that these models were limited in their ability to identify the social actors that were at the heart of these strikes. To address this limitation, he began to use newspapers as a new data source for identifying social actors. While he was conducting his analysis, it became clear to him that all narratives have an invariant structure, which includes a subject (social actor), a verb (social action), and an object. According to Franzosi, this structure (SVO) is rooted in rhetoric (the study of persuasive communication), which he explained was first introduced in ancient Greece, and later translated by Thomas Wilson in *The Arte of Rhetorique*[1] to what are now often referred to as the five W's of journalism (who, what, where, when, and why).

Franzosi pointed out that because of the immense volume of data available today, computational methods are a necessity in narrative research. Therefore, he developed software that would allow him to identify the SVO structure in narratives quickly and accurately using a method he terms "quantitative narrative analysis." With this software, Program for the Computer-Assisted Coding of Events, he was able to examine approximately 50,000 newspaper articles from 1919 to 1922, which resulted in the identification of 250,000 SVO sequence sets related to the rise of Italian fascism in postwar Italy.

Returning to the five W's structure, Franzosi observed that while the SVO structure addresses the questions who and what, geographic information system (GIS) models are helpful in identifying where actors act. As an example, he referred to his research on the emergence of Italian fascism, in which he used a GIS model to track the activity of socialists and fascists from 1919 to 1922. He found that the location of socialists leading the revolutionist movement from 1919 to 1920 overlapped with the locations in which fascism first emerged in 1921 and 1922.

Franzosi closed by sharing two different approaches currently being used by researchers to extract the SVO structure in narrative analysis. He first described ClausIE,[2] German open-source freeware software based on the Stanford CoreNLP natural language software.[3] This online open information extractor uses automated methods to identify the SVO structure in text. Franzosi characterized the second approach as similar to novelist Kurt Vonnegut's "man in a hole" narrative. Vonnegut posited, Franzosi explained, that stories often follow specific patterns such as "man in a hole." In this narrative, a character who is relatively happy receives bad news (or something bad happens to him). This bad news causes a temporary dip in the man's state of mind or welfare until his luck inevitably changes, and he receives good news (or something good happens to him), which pulls him out of the "hole." According to Franzosi, researchers are now able to plot these arcs in narrative by using sentiment analysis (computational analysis used to identify the intended sentiment for a set of words) and complex

---

[1] Wilson, T. (2010). *The Arte of Rhetorique (1560)*. Oxford, UK: Benediction Classics.

[2] For more information on ClausIE, see https://www.mpi-inf.mpg.de/departments/databases-and-information-systems/software/clausie [April 2018].

[3] For more information on Stanford CoreNLP, see https://stanfordnlp.github.io/CoreNLP [April 2018].

algebraic matrix factorizations, such as singular-value decomposition and non-negative matrix factorization.

## ADVANCING NARRATIVE RESEARCH WITH BIG DATA SCIENCE

Mark Turner, Case Western Reserve University, first conceived the idea of building a data repository for the study of communication in the early 2000s. He had observed large laboratories in the genomic and other biological sciences working together to build large online databases to store and share data collectively, and said he hoped to provide a similar resource for studying different modes of communication. In 2010, he and a colleague developed Red Hen Lab, a global consortium for communication research.[4] According to Turner, the repository currently stores approximately 4 billion words and 360,000 recorded hours of audiovisual (AV) broadcasts that are searchable using natural language processing (NLP) tagging tools, such as Stanford CoreNLP and Apache OpenNLP, and optical character recognition software, such as Tesseract. The most common type of data contained in the repository, he noted, is AV news broadcasts in such languages as English, German, Portuguese, Russian, Czech, Arabic, and Chinese.

Turner explained that, although the main goal of the consortium is the development of theory, he and his colleagues are also interested in developing the computational methods needed to conduct theory-based research. According to Turner, for example, the repository currently has the ability to identify and tag frames—cognitive structures that determine the process and result of interpreting linguistic forms, such as words, phrases, and grammatical patterns[5]—so that they can be searched along with grammar and construction and word strings. To illustrate the use of this method, he cited its ability to help researchers identify locations where there is an emerging risk to a region's stability because of such significant events as extreme weather. He added that the consortium also benefits from the expertise of researchers and students in other disciplines, including linguistics and machine learning, who share with the consortium methods and techniques that further advance the study of narrative. Turner closed by noting that these methods, combined with the consortium's data repository, enable researchers from around the world to track and study narratives in real time.

---

[4] For more information on Red Hen Lab, see http://www.redhenlab.org [April 2018].
[5] Heine, B., and Narrog, H. (2009). *The Oxford Handbook of Linguistic Analysis*. Oxford, UK: Oxford University Press.

## FUNCTION WORDS AND THE ARC OF NARRATIVE

James Pennebaker, University of Texas, Austin, examines narrative from a psychological perspective. He explained that he was first introduced to narrative research while conducting an experiment on expressive writing. It is well known in medicine and psychology, he observed, that individuals who have experienced a major traumatic event at some point in their life are much more likely to develop health problems, and the likelihood increases if their traumatic experience has been kept secret. Speculating on the effect writing about these secret traumatic experiences might have on a patient's health, he conducted a study in which participants were randomly assigned to write about either a personal trauma or a less personal, superficial experience. He and his colleagues then monitored the participants' health using such markers as immune function. The results, he reported, showed that the act of writing about a personal trauma can produce positive changes in physical health.

Pennebaker and his students then developed a software program that would allow them to analyze text more efficiently. This word counting software program, he explained, known as Linguistic Inquiry and Word Count (LIWC), made it possible to analyze text in a new way, providing the ability to calculate the percentage of positive and negative emotional words; cognitive process words, such as "understand," "realize," "know," or "meaning"; and parts of speech, such as pronouns, prepositions, articles, conjunctions, and auxiliary verbs.

Using this computational analysis, Pennebaker learned that the way people tell a story can often be more revealing than the story's content. By looking at function words (e.g., pronouns, prepositions, articles, conjunctions, auxiliary verbs),[6] for example, he and his colleagues were able to identify a surprising number of characteristics of the authors of the stories being analyzed, including their genders, intelligence levels, emotional states, and social connections. The use of function words, they found, could reveal whether the author was being honest or not.

Intrigued by these results, Pennebaker conducted additional studies using LIWC analysis. In one study, he analyzed stories created by students after they had participated in a thematic apperception test.[7] The study revealed, he reported, that graphs showing the number of function words used in a story typically reveal a particular pattern: they are used at a high

---

[6] Pennebaker explained that function words are social words that are processed differently in the brain than are content words, which include nouns, regular verbs, most adjectives, and most adverbs.

[7] A thematic apperception test is a psychological test used to reveal thoughts and attitude patterns by examining the stories people devise when presented with a picture involving people unfamiliar to them. For more information, see http://www.utpsyc.org/TATintro [April 2018].

rate in the beginning and then eventually drop off, creating an arc. In what became known as the "arc of narrative" project, he analyzed thousands of novels, short stories, Supreme Court decisions, and movie scripts. The arc of narrative pattern, he discovered, can be predictive of positive emotional states and successful outcomes. For example, he and his colleagues analyzed stories that detailed romantic breakups. They found that the arc of narrative pattern was present in such stories told by authors who had recovered from the event but not in those told by authors who had not yet recovered. Similarly, they found that movie scripts containing the arc of narrative pattern had higher ratings than those that did not.

Pennebaker closed by observing that he and his colleagues are currently analyzing stories told by members of ISIS, as well as ones already identified as having been told by liars and truth-tellers. Such research, he explained, is allowing researchers to gain understanding of how people and groups construct their stories and histories.

## THE SMALL-STORY APPROACH TO NARRATIVE RESEARCH

Michael Bamberg, Clark University, divides narrative analysis into two categories: big-story and small-story approaches. Big stories, he said, help people make sense of the world through the narratives told by nations, organizations, institutions, and individuals, citing the example of the story of how the allies sought to contain the spread of communism after World War II. He defined small stories, on the other hand, as everyday stories such as fairytales, novels, and personal conversations. Narrative analysis, he explained, typically focuses on the big stories and addresses the structural, textual, and thematic aspects of their narratives, whereas he is more interested in when, why, and how small stories influence big stories.

Bamberg described research he conducted based on the hypothesis that one way small stories differ from big stories is in their formation. He noted that while some small stories, such as novels and fairytales, are presented in familiar structures, conversational small stories are not, and because they are regarded as mundane, they are often overlooked. He explained that his small-story approach to narrative analysis examines how a story originates, is picked up by others, and later transforms over time. He added that he used several methods common in communications research, such as conversation analysis of both verbal and nonverbal communication. He noted that gestures, posture, and facial expressions can often reveal how a story is received and understood by others. He suggested, however, that more research is needed on visual narratives, such as those used in commercials to facilitate highly emotional plot structures.

In closing, Bamberg pointed out that while small stories are often formed and shaped by big stories, it is also likely that big stories form from small stories. Similarly, he said, it is possible for small stories to challenge or change big stories. He concluded by suggesting that research to improve understanding of these two phenomena would be beneficial for intelligence analysis.

## DISCUSSION

Turner opened the discussion by agreeing with Bamberg that research on visual narratives is needed. He noted that he and his colleagues at the consortium have been collecting and tagging a variety of visual data to store in the Red Hen Lab repository, adding that even subtle facial expressions and gestures can be detected and tagged so that researchers have access to all forms of communication.

The remainder of the discussion focused on comparing methodological approaches and considering how narrative research might help analysts determine whether a narrative is true or false and whether it is changing at the societal level.

One participant commented on the range of approaches presented by the speakers. Franzosi and Pennebaker, he suggested, focus their analysis on text and sentence structure, while Bamberg's approach is more qualitative. The consortium for data science discussed in Turner's presentation sits somewhere between these two approaches, he added. Pennebaker responded that both qualitative and quantitative approaches are necessary for narrative research. Elaborating, he explained that although the quantitative methods used in his work are effective and accurate, they are complemented by the more individualized qualitative methods of studying narrative. Turner was asked whether it is possible for big data researchers to understand their data as well as qualitative researchers working in the field. Turner asserted that, while qualitative researchers are more familiar with their data, computational and statistical methods may provide insights not available through qualitative methods. He added that members of the consortium plan to continue collecting data that are multimodal and current so analysts can observe narratives as they evolve in real time.

Responding to a similar comment on the importance of measuring outcomes, Pennebaker agreed and suggested the need for a shift from the use of traditional outcome questionnaires to measuring behavior to determine how narrative impacts a person's life. He commented that the research discussed by Franzosi is important because, by revealing that the manipulation of language can cause changes in a person's emotional state, it revealed a connection between language and behavior.

Turning to the topic of rhetoric, one participant noted that analysts are often asked to make sense of a situation on the basis of limited information. However, since narratives are designed to be persuasive to both the person constructing the narrative and others, he wondered whether the result could be analyses that are more imagination than reality. Another participant called attention to the significant body of research in narrative theory on fictionality, which separates the idea of general fiction (e.g., novels, films) from the idea of fictionality as a method of invention.

Noting that analysts are asked to make predictions on both the strategic and tactical levels, one participant asked how a research model might reveal when a phenomenon such as preference falsification (when people are expressing what they consider to be socially acceptable preferences rather than their true preferences) is occurring. If an analyst is aware only of the societal narratives created by preference falsification, she added, it may be more difficult to predict when social change is about to occur. Pennebaker explained that studying the use of function words can reveal things the speaker or author may be trying to hide with the use of content words. For example, he said, by studying George W. Bush's speeches, he could identify when Bush decided to enter the war in Iraq. Approximately 9 months before doing so, he elaborated, Bush significantly decreased his use of the pronoun "I." According to Pennebaker, use of the word "I" indicates when someone is being personal. Only after the United States entered Iraq, he added, did Bush return to using the word "I." Pennebaker also noted that the Boston Marathon bomber stopped using the word "I" in his social media communications before the bombing.

# 3

# The Science of Narrative Communication

The second session of the workshop was designed to provide an opportunity to explore the potential advantages of the study of narrative for national security. Three panelists made presentations about relevant research and pointed to methods for drawing meaning from narrative analysis, which set the stage for an open discussion among participants.

## NARRATIVE PROCESSING AND SENSEMAKING

Michael Dahlstrom, Iowa State University, drew on his work in the study of science communication to explore the role of narrative in how people come to understand and act on science information. He noted that there are three distinct concepts of what narrative can be, and that distinguishing among them helps to clarify discussions about narrative influence.

The first concept of narrative, Dahlstrom explained, is a form of information processing, which is one of two contrasting pathways through which information is processed: narrative and scientific. Because humans are natural storytellers, he elaborated, many scholars assert that humans process information predominantly through narrative pathways, which he characterized as a natural, efficient, and easy means of information processing. In contrast, he continued, scientific processing is more challenging and thus takes effort, requiring analytical thinking skills to process facts and evidence. Dahsltrom cited controversy over the safety of vaccines as an example of how these two pathways compete. Vaccine proponents, he explained, often rely on scientific evidence to communicate the message that vaccines are a safe and effective way to prevent disease, whereas members of the antivaccine movement often share stories about dangerous side effects experienced by vaccinated children, which are processed as natural narratives.

A second concept of narrative, Dahlstrom continued, are sensemaking narratives, also referred to as internal narrative frameworks. These are individualized stories of cause-and-effect relationships that emerge in the mind from direct and/or mediated experience and are often the product of processing information through the narrative pathway, he explained. These sensemaking narratives accumulate over time and create a foundation used to make sense of the world. Dahlstrom added that sensemaking narratives also guide what information people seek out and how, and affect how they interpret newly received information.

Dahlstrom then turned to the third concept of narrative—external narrative messages—which includes stories people receive from outside sources, as well as stories they share with others. Not all messages are narratives, he noted. The amount of narrativity (the characteristics that make up a narrative) contained in a message varies, he said, and he explained that studies on narrative persuasion (the persuasiveness of stories) and narrative transportation (immersion into a narrative) have found that external messages with more narrativity are often more persuasive and engaging.[1] However, he added, because narrative messages are intrinsically persuasive, even those containing inaccurate scientific information can still influence others. And while it may be assumed that inaccuracies can be corrected by providing more scientific evidence, he observed, communication researchers have learned that this is rarely successful because the existing sensemaking narratives will influence how that evidence is interpreted. For elaboration of this idea, Dahlstrom pointed to research on cultural cognition by Dan Kahan suggesting that even individuals with accurate knowledge about scientific issues can become polarized.[2] When this polarization occurs, he explained, any additional facts received will be interpreted based on existing frameworks and used to support existing sensemaking narratives.

In closing, Dahlstrom stated that narrative pathways, sensemaking narratives, and external narrative messages work together to create a "symbolic reality" for individuals or for groups that share certain characteristics or life experiences. He suggested that research into how this symbolic reality moves and shifts in relation to the international landscape could be use-

---

[1] Dahlstrom, M.F. (2014). Using narratives and storytelling to communicate science with nonexpert audiences. *Proceedings of the National Academy of Sciences of the United States of America, 111,* 13614–13620.

[2] For more information on Dan Kahan's Cultural Cognition Project, see http://www.culturalcognition.net/kahan [April 2018].

ful for making analytical predictions because, he said, people do not make decisions based on reality—they make decisions based on the narratives that create their symbolic reality.

## NARRATIVE, CULTURE, AND DIGITAL TECHNOLOGY

Pauline Cheong, Arizona State University, drew on her work on the nexus of culture and communication technologies to consider why the study of narrative is important. Humans are "narrative beings," she observed, and "we make sense of our role in large part through the stories we know are truth and share." She added that narratives are important because they help shape people's ideas and interpretation of events.

Cheong went on to identify storytelling as a crucial part of cultural reproduction—the continuation of a culture across generations—but suggested that its nature is changing in response to the development of new digital technologies. Storytellers, she said, are finding new ways of voicing and composing stories as digital technologies continue to evolve. As a communications researcher, she studies how narratives are formed and spread across online platforms using such theories as that of "convergence culture," which posits that the convergence of old and new media facilitates transmedia storytelling (telling stories across multiple platforms using digital technologies) to audiences that have become accustomed to seeking and connecting information as it flows across media platforms.[3]

Although research on transmedia storytelling has traditionally focused on stories designed for entertainment, such as novels and movies, Cheong is interested in how the study of narrative and transmedia storytelling could be applied to understanding such complex and strategic issues as national security, religious authority, and community. This type of research, she continued, is essential to intelligence analysis because narratives, which help shape ideology and the interpretation of events, have a strong relationship to knowledge and power. She suggested that such analysis is especially important for understanding hegemonic struggles, where the meaning of truth becomes a key feature in the symbolic battle for hearts and minds. Changing media conditions, she said, allow nonstate actors to communicate their practices and tactics of resistance at a reasonably low cost, thereby disrupting the strategic communications of nation states and in turn subverting state ideology and national branding. She suggested that research should focus on how the changing multidimensional nature of strategic communications influences the perceptions and credibility of political and religious leaders in struggles for authority and power.

---

[3] For more information about Henry Jenkins' theory of convergence culture, see http://henryjenkins.org/blog/2006/06/welcome_to_convergence_culture.html [April 2018].

In *Narrative Landmines: Rumors, Islamic Extremism, and the Struggle for Influence*,[4] Cheong and her colleagues suggest that "narrative offers a means of uniting culturally-provided templates," which include histories, rumors, and other story forms. Thus, she explained, truth becomes less about facts and evidence and more about narrative fidelity (how well a story resonates with listeners as a result of their experiences and beliefs). For example, she noted, media platforms can be used by nonstate actors to portray terrorists as either heroes or outlaws.

Depending on the cultural context, Cheong continued, open-source narratives such as these have the power to facilitate middle-ground resistance among civilians. It is possible for researchers to trace the formation and evolution of these narratives, she stated, by looking at how they are received and shared online. Thus, she suggested, research on open-source and online narratives could supplement actionable intelligence and give governing leaders the ability to counter false messages, shape perceptions, and mobilize against threats.

## COMMUNICATING AND SENSEMAKING THROUGH COMPRESSED NARRATIVES

Humans are unique in their ability to think across causation, agency, time, and space, observed Mark Turner, Case Western Reserve University, in explaining why understanding narratives is a vitally important kind of analysis. It is with narrative ability, he continued, that humans can think about and plan for the possibility of future events, such as international state stability. To make vast and complex narratives more digestible, he explained, they can be compressed and stored in the brain for future use, and when a person encounters new data or facts, these compressed narratives can then be used to aid in the sensemaking process. The process continues, he said, as newly formed compressions are then stored in the brain or blended with other narratives.

To clarify how narrative compressions are formed, Turner shared a promotional video for China's Belt and Road Initiative.[5] The video, which begins with images of footsteps from the past that transform into images of transportation and trade opportunities, is a compression of narratives visualizing the importance of trade in China's past and future.

---

[4] Bernardi, D.L., Cheong, P.H., Lundry, C., and Ruston, S.W. (2012). *Narrative Landmines: Rumors, Islamist Extremism, and the Struggle for Strategic Influence.* Piscataway, NJ: Rutgers University Press.

[5] For more information on China's Belt and Road Initiative, see http://english.gov.cn/belt AndRoad [April 2018].

Turner identified several research questions that could advance the field of narrative:

- What patterns of compression are normal?
- What kinds of patterns of compression work?
- Which patterns of compression are cross-cultural?
- Which patters of compression are culture-specific?
- What are the mental operations for making compressions into a narrative?
- What are the mental operations for decompressing narratives to connect with other narratives?

## DISCUSSION

The open discussion following the presentations previously summarized began with thoughts from each of the panelists about how to define narrative. Noting that the words "story" and "narrative" appear to be used interchangeably, moderator Matsumoto also asked the panelists to clarify differences between the two terms.

Dahlstrom's view was that there could be several ways to define the term "narrative." At the most basic level, he stated, a narrative occurs when something happens to a character of some kind (not necessarily a human being), and some sort of change occurs. Beyond that, he said, there is "a huge range of what narratives turn into." He explained that narratives based on personal experience are used for "internal sensemaking" and may be used only by the individual, whereas narratives that are shared with others in some way have a purpose. In general, he continued, for a storyteller to consider a narrative to be worthy of telling, it must conflict with one of the storyteller's normative expectations of reality: if there is no reason to tell the story, it will seem pointless to the hearer. Thus, he observed, knowing which normative expectation was broken can offer insights into a person's expectations and perceptions about the world.

In contrast, Cheong explained that she focuses not on stories themselves but on how narratives are embedded in a particular context, which "provides the backdrop for understanding and interpreting the meanings or multiple layers of meaning" they contain. She and her colleagues have defined narrative as a "system of interrelated stories that share common elements and a rhetorical desire to resolve a conflict by structuring ordinance, expectations, and understandings."[6] She pointed out that the 30-second

---

[6] Bernardi, D.L., Cheong, P.H., Lundry, C., and Ruston, S.W. (2012). *Narrative Landmines: Rumors, Islamist Extremism, and the Struggle for Strategic Influence*. Piscataway, NJ: Rutgers University Press.

trailer about China presented by Turner illustrates the need for multilayered understanding of a narrative's context because the narrative could be interpreted in many ways, and discerning which interpretation may be accurate would require understanding the purposes for which it was developed.

Medina noted that she is often intrigued by the statements made in online comment sections, such as those found on YouTube. She wondered whether Cheong considered these to be relevant when collecting online data. Cheong agreed that online comments present a promising research opportunity. Although researchers often consider them to be trivial, she continued, they are a part of the narrative and thus a valuable source of data.

Referring to the "system of interrelated stories" included in Cheong's definition of narrative, Medina asked whether she agreed that nation societies differ in their ability to project and control narrative. Cheong identified this behavior as "nation branding," and explained that these are stories nations tell about their origins and development as a way to promote their country and strengthen soft power. She agreed with Medina's assessment that some countries are better than others at nation branding, citing South Korea as having become particularly savvy in this regard. In recent years, she continued, South Koreans have begun to shape their national brand with compelling narratives dispersed via entertainment platforms and popular culture.

Turner added that narratives can also be crafted to appeal to certain audiences. For example, he said, the promotional video he used to demonstrate compressed narratives was prepared for English-speaking audiences, and the message would likely have been crafted differently if it had been meant to appeal to another culture. Dahlstrom pointed out that mass media also play a role in the decline of shared narratives. He explained that people are beginning to use the Internet as a way of connecting with others who share views that may differ from a culture's shared narratives. He suggested that research opportunities exist in learning when and how ideas separate and what interventions might be developed to counter groups that may spread harmful messages.

One participant pointed out that by using the structure of cause and effect, narrative targets emotions. Cheong agreed and suggested that emotion is important for narrative. However, she added, only recently have communication researchers begun to recognize the significance of emotive language and consider which methods might be used to analyze it.

Another participant noted that although narrative helps people make sense of the reality in which they exist, it is not reality. He wondered, then, whether by stepping into the narrative domain to help others make sense of the world, researchers and analysts are at risk of going off track. He suggested that this may be even more of a concern when working with big data. Another obstacle, Medina added, is that the Intelligence Community also has a narrative that it uses to make sense of reality and other narratives. Cheong suggested that the tools discussed in the first workshop panel (see Chapter 2), such as coding data and identifying sentence structure, might help with addressing this issue. However, she cautioned, because these tools are based on rational discourse, they may not capture emotive language and disruptive storytelling.

One participant asked how to communicate understandable narratives related to complex scientific subjects for audiences that may include people who do not understand the data. Dahlstrom suggested that the answer depends on the goal: if the goal is to develop analytical thinking, narrative processing methods will not be effective, whereas if the goal is to increase understanding, it will be wise to concentrate on creating a narrative that expresses the information. He added that the person communicating this message must also decide whether the goal of the narrative is to persuade listeners or to provide information in a way that allows them to form their own opinion.

Another participant noted that many people are susceptible to believing false narratives and asked the panelists whether they could suggest methods that might inoculate the public against such narratives. Dahlstrom acknowledged that, while he does not know the answer to this question, he does know that adding more facts to a situation will not solve the problem. However, he added, crafting a competing narrative may backfire and instead cause the audience to become skeptical, which in turn will likely make the false narrative more appealing. He suggested that technological advances may help identify when mental narratives are forming and when they may be shifting away from the scientific message. He argued that such insights might make it easier to identify where opposing beliefs connect and how they could be merged.

Medina questioned whether narratives can be studied in real time. Turner explained that big data science tries to do this by collecting information consistently across media platforms. He added that huge repositories of data allow researchers to track events as they occur and spread throughout other countries and cultures.

# 4

# Narrative and Emerging Technologies

The third session of the workshop turned to the ways artificial intelligence (AI), big data, and other emerging technologies may be influencing the nature of narratives and the ways in which they are formed, as well as the impact of social media on the development and dissemination of narratives. Moderator Doug Randall, Protagonist, opened the discussion with a brief look at some of the ways technology is affecting daily life. The average American, he noted, spends 12 hours per day consuming electronic media of some sort, including 2 hours spent on computers, 3 on mobile devices, and 4 watching television. "Those are a lot of narratives getting thrown at us every day," he commented, adding that these interactions produce an "overwhelming amount of data," as well as an opportunity to understand developing narratives. In his view, rapidly increasing rates of both financial investment and research in developing technologies, including AI, machine learning, natural language processing, and social media monitoring, reflect the major impact these technologies are having.

### COMPUTATIONAL MODELING OF NARRATIVES

A computer scientist by training, Michael Young, University of Utah, described aspects of his work in building computational models that can be used analytically. He began by explaining that his objective in building computational models of narrative is to produce not only textual narratives but also cinematic narratives for 3D virtual worlds and interactive narratives for virtual environments and computer games. He noted that his work draws on collaborations with cognitive psychologists who have studied how people develop mental models, some of which was originally developed for modeling and programming robots.

Young's first step in generating a narrative model is to form representations using large amounts of event data. He then creates a structure for the story and the underlying discourse to convey the story through text or cinematic means. It is this first step, he noted, that is most relevant to the Intelligence Community (IC), as the representations used to produce the narrative could help an analyst make sense of the data. He added that the storylines are based on what he termed primary or atomic building blocks—events or actions that propel a story forward—and the algorithm he develops to model the story sequences the blocks and searches for all possible threads that might connect them.

Young explained that when building narrative models, he is guided by research from cognitive psychologists on mental models (conceptual representations used to make sense of the world). Mental models, he continued, tend to remain constant across media and various interactions, so that when situations are created as they progress within the telling of the story, causation, intentionality, and actions can be based on these mental models.

## ATTITUDES, BEHAVIOR, AND INFLUENCE

Catherine Tejeda, Parenthetic, offered brief remarks on her research on the question of how human–computer interactions influence the flow of narrative. She is particularly interested in the methods used by marketers to measure attitudes and behaviors, which are in turn used to determine what information is directed to individuals and groups online and how. One way marketers measure attitudes and behaviors, she explained, is by observing the choices a person makes online. Such actions as shopping online, reading news articles, and posting on social media can reveal a person's attitudes and behaviors, she added. Just as nations are branded by the image they choose to portray, she suggested, online actions determine how a person is perceived by those observing their behavior.

## DISCUSSION

Questions raised in the discussion session focused first on the context for understanding narratives. Referring back to earlier discussion of the point that narratives exist and are understood within the scope of narrative landscapes, a participant asked whether there is a similar landscape that must be considered when building a model and what this means for the creation of models. Young responded that this is actually a critical perspective in computational research and that a debate exists as to whether story structure is separate from the discourse component. For example, he elaborated, if an engineer creates a story, some believe that it should then move to another department so that someone else can work on the discourse. Young believes that the communicative goals of the story can best be achieved if the story and discourse are created concurrently.

Another question raised was about whether the link between expressed narrative (purposefully influencing a narrative to affect public perception) and behavior is a consideration in the computational modeling of narrative. Young responded that, unlike synthetic narratives (narratives automatically fabricated by a software program), the narratives generated from existing data must be created by the actions of a participant (e.g., a game player or social media user). However, he continued, while participants do influence events, they do not act with the intention of influencing the narrative. In fact, he explained, participants are usually operating with limited knowledge of the environment in which they are acting.

Participants also raised questions related to AI. Randall wondered what advances in this technology might mean for people's ability to analyze, understand, and shape narratives in the future. Tejeda predicted that the use of automated conversation, which is based on machine learning, is likely to increase in the future. Although it is used primarily as a way to initiate chats with online customer service agents, she observed, it has recently become popular with health firms, medical practices, and large health insurance agencies. Young pointed out that while machine learning has experienced great successes in recent years, it is only one form of AI. He suggested that the combination of increased access to data, improved processing power, and new algorithms has allowed researchers to extract data structures that have helped solve critical engineering problems. However, he asserted, while machine learning will most likely continue to advance such scientific breakthroughs as self-driving cars, it will not answer such important questions as how to understand the cognitive processes around narrative.

Making reference to a science fiction–based article that describes a machine so advanced that it sought to unite the world by releasing emotionally compelling stories on a global scale, one participant wondered whether the panelists believed something like this could ever happen. Young suggested that the machine itself would not have the power to do such a thing. An argument could be made in terms of the number of people that received the narrative, but the power would come from the narrative rather than the machine. A participant suggested that, while the infrastructure for such an event exists, the real issue when thinking about the needs of the IC would be how best to time the release of such an event (e.g., gradual release, over a period of 5 years, immediately for a more disruptive event).

Another participant suggested that persuasion typically occurs at the individual level through one-on-one conversation. Scientists, he continued, are beginning to develop machines to initiate conversation for the benefit of humans. He suggested that it may not be long before robots can be used to create and shape narratives so compelling that humans will become engaged in back-and-forth conversation with the robots. He was curious to know whether the panelists had encountered this type of human–machine interaction in their work. Young replied that, although he does not work with robots, he does see this type of interaction in computer games. He illustrated the point by explaining that when people begin playing a computer game, they enter a virtual world that requires them to make choices that determine how a story will unfold. If a player unintentionally makes a choice that derails the storyline, the computer will respond by creating a new pathway so that the player can continue to move through a new narrative arc without ever knowing a derailment occurred. Young noted that this technology is also used to create training programs, such as those used to build social skills in soldiers. The interactive narrative generation allows the computer to personalize the training to the individual. Sara Cobb, George Mason University, added that, just as new structure patterns could be created in Young's example, it may be possible to design bots that can create structure patterns to encourage positive outcomes for real-world issues, such as conflict deescalation, health, positive emotion, and prosocial values.

Referring to Tejeda's presentation, a participant asked how influence is related to such technologies as machine learning and AI. Returning to the subject of measuring attitude and behavior, Tejeda responded that technology helps researchers move beyond traditional survey work into more observational techniques. For example, she said, in addition to "clicks" and "likes," a researcher might analyze online comments to provide context for those choices. Young pointed out that technology could also be used to influence the way stories are perceived and understood. For example, he explained, a computational model might influence participatory thinking by generating a storyline that is presented in a way that builds suspense. It is also possible, he suggested, that a storyline could be presented to an analyst in a way that would influence that person to highlight certain inferences.

Thinking in terms of the spread of online messages, one participant asked whether most people are more concerned with the number of times something is shared (e.g., Kim Kardashian shares content with her online followers) or how it is shared (e.g., Kim Kardashian shares content with her followers and states that it changed her life). Tejeda responded that there is usually more concern as to how content is shared. Randall followed up by noting that the extent to which people engage with the content and how it is referenced are also important considerations.

Another measure of interest in narrative research, Randall suggested, is engagement. He argued that measuring both the metadata and the substance of communications can help researchers know whether people are accepting and connecting with the narrative. Young observed that computer scientists tend to be more concerned with comprehension than engagement. They also want to know, he continued, how the structures used to increase comprehension contribute to engagement or to proxies for engagement, such as suspense. "What is the cognitive machinery around suspense?" he asked rhetorically. "Is that knowable or can we begin to view the cognitive processes like engagement as a result of the way that we design the narrative itself?" Tejeda commented that reach is also important when measuring engagement—in other words, not just who engaged with the content but who saw it.

A participant noted that, if combined, new technologies such as IBM's Watson[1] (a cognitive computing system) and Cambridge Analytica[2] (a data mining and analysis program capable of public profiling) could have the power to exert influence on a scale that would be both marvelous and terrifying. The participant added that because such technological advances are making it easier to manipulate narrative, knowing how narratives are created, manipulated, and disseminated could help protect people from toxic narratives. Tejeda commented that while the possibility of companies using technology to influence the choices made by an individual is disturbing, she is not sure that a mass influence campaign would be effective. Returning to the idea that individual narratives are understood and shaped by narrative landscapes in which they fall, she suggested that attempts to influence on such a large scale may fail.

Thinking about the power and scope of narrative, Matsumoto asserted that learning how a narrative might be limited or stopped is also essential to understanding influence. Furthermore, he asked, "What is the source that is going to stop it? Is that source available right now so that it limits the everyday influence of narrative?" Randall suggested that while technology is advancing and researchers are working to collect large amounts of data, the field is limited by the fact that it is not there yet. Johnson made the point that not all narratives are believable. Tejeda agreed and added that it is not always easy to trust that a source is providing accurate information. Young suggested that the way a text is written may play a role in its believability, with some narratives perhaps being easier to follow than others. Furthermore, he noted, as had been mentioned in previous panels, the narrative may not line up with a person's internal narratives. Remarking that work on credibility indicators has been done in other domains, Matsumoto wondered whether the panelists were aware of similar research done to index and measure believability indicators. Young acknowledged

---

[1] For more information on Watson, see https://www.ibm.com/watson [April 2018].
[2] For more information on Cambridge Analytica, see https://cambridgeanalytica.org [April 2018].

that researchers in computational modeling are beginning to look at these questions, but they are not far enough along to model them.

In the social sciences, Roberto Franzosi, Emory University, explained, researchers use theory to develop models that can then be used to make predictions. However, he continued, there has been a recent trend among computer scientists to suggest that big data has made theory obsolete. Young agreed that this is a problem. While the data-driven approach may work well for solving engineering problems, he observed, very few scientific questions are answered this way. For example, he said, if one is studying the interactions among invariant properties in an external model (e.g., cognitive processes), an internal model (e.g., the structural processes within the properties of a narrative), and an artifact (the interface or discourse) to see how they might change when manipulated, theory is needed to answer the questions that arise.

# 5

# Narrative and Power

The fourth panel of the workshop, moderated by Jeffrey Johnson, University of Florida, and Sara Cobb, George Mason University, focused on the relationship between narrative and power. Johnson opened the session by noting that it would address such questions as (1) how narratives vary across cultures and how those narratives clash, (2) what the outcomes of clashing narratives might be, and (3) how narrative might be used as a tool for mobilizing intervention.

### COMPETING NARRATIVES

Narrative is a way of organizing, giving meaning to, and creating understanding of our experiences, explained James Phelan of Ohio State University. His rhetorical definition of narrative—somebody telling somebody else on some occasion and for some purpose(s) that something happened—emphasizes the importance of tellers, audiences, and purposes. The teller gives shape to the raw material underlying the narrative (primarily events and characters in space and over time) by choosing some techniques rather than others, some way of arranging and emphasizing aspects of that raw material. The teller makes those shaping choices in light of his or her purpose and audiences. In any nonfiction narrative, Phelan noted, the raw material exists independently of its treatment, and thus it can be shaped in different ways by different tellers. Thus, any nonfiction narrative can be contested by one or more others.

Phelan went on to identify a number of variables used to adjudicate competing narratives. One is an appeal to the phenomena being explained.

As an example, Phelan cited raw material that is captured in a way that is coherent and precise so that it appeals to logic. He also gave the example of an appeal based on the position and perspective of the teller or audience, noting that the teller is likely to privilege his or her particular point of view, and the audience can either accept or reject the narrative. He pointed to relative power as another relevant factor, explaining that someone with power will present a narrative aimed at preserving that power, while someone without power can use narrative as a way of gaining power. Knowing and understanding the source of the narrative is also key, he suggested, noting that the reputation, trustworthiness, and credibility of the teller are all considerations in assessing the narrative. He added, however, that, regardless of the source's credibility and trustworthiness, the narrative must also serve the audience's interests for them to accept it. Finally, he pointed out that for one narrative to win over another, it must be disseminated to reach the desired audience at the appropriate time.

Phelan also discussed the role of narrative in mobilization and intervention. Using fiction as an example, he introduced the audience to what he called mimetic and thematic links. In fiction, he explained, a character can be thought of as having both mimetic and thematic components. The mimetic component identifies the way a character is a possible person, and the thematic component identifies the way a character is a representative figure (of a group and/or one or more ideas). According to Phelan, effective narratives make a seamless connection between their mimetic and thematic components, and in so doing, "marshal cognition, affect, and values in the service of an idea or certain positions."

A major source of conflict between narratives is difference in cultural values, Phelan explained. Cultural values, he elaborated, reflect the organizing schemas used to understand and construct hierarchies among experiences within a culture (e.g., good vs. bad, right vs. wrong, better vs. worse). He highlighted the recent #MeToo movement as an example, suggesting that the success of the movement both reveals and contributes to a shift in cultural values. Although no one before the movement would have openly argued that sexual misconduct by powerful men against less powerful women was a good thing, he said, the culture has shifted to openly denouncing such behavior. His last point was that moving beyond the conflicts that arise with competing narratives requires some recognition of common ground or agreement on the part of those involved.

## COUNTERING TOXIC NARRATIVES

As a marketer, Debra Louison Lavoy, Narrative Builders, routinely works with public and private organizations to develop influential narratives. She defined narrative as "an interconnected set of beliefs that influence the way you interpret the meaning of things." Narratives are very powerful, she asserted, because once a narrative has been internalized by the audience, it can affect how people see the world and their attitudes and behaviors.

Louison Lavoy went on to explain that digital marketers have learned two ways to wield power: by constructing narratives and by disseminating narratives via digital channels so that others can be persuaded by them. She has developed a heuristic that includes five measures—presentation, clarity, resonance, sharability, and organization—for evaluating the power of a narrative. She then expounded on each of these measures in turn:

1. Presentation includes such visual and auditory components as music, word choice, and font. Furthermore, because it is important to be able to reach and engage a wide audience, addressing such issues as accessibility is also important.
2. Clarity is a measure of how well the narrative message is received by the audience: "Can I understand your story, your narrative once you have explained it to me?"
3. Resonance includes emotional resonance (connecting to emotions that matter to the listener), intellectual resonance (credibility), and echo (whether the narrative is similar to others that are already part of the psyche).
4. Given the increasing importance of social media, sharability may be the most important measure, Louison Lavoy noted. She cited linking the narrative to other content and actively encouraging the audience to share the narrative with others as the two most important components for sharability.
5. Finally, organization refers to the structure of the narrative. Louison Lavoy explained that a structured narrative should first address why it matters, adding that the teller must also consider how to convey a particular vision to the audience in a way that makes sense to them. Another component of organizing a narrative for public consumption is what she called the "offer" or "ask." For example, she said, nongovernmental organizations and politicians will often ask for volunteers, donations, or votes.

Louison Lavoy added to this list that testimonials, data, and other forms of evidence can be proof that a message is credible and not "too good to be true."

Louison Lavoy closed with the point that while narratives can share positive messages, they can also be used to spread toxic ones. A narrative is considered toxic, she explained, when it is "intentionally based on false or misleading information." To counteract toxic messages, she suggested that researchers consider studying how marketing strategies might be incorporated into social science theory to (1) identify and block toxic campaigns, (2) inoculate populations against them, and (3) develop countermessages.

## THE POWER OF NARRATIVE

Michael Bamberg, Clark University, began by asserting that when talking about the power of narrative, one should also talk about narrative emotion. He explained that while narrative emotion is typically studied by looking at text arrangement or character positioning, researchers can also examine how plot is used. Approximately 32 forms of plot exist, he added, each of which uses a different approach to emotional engagement.

The power of narrative can also be studied by observing and listening to speakers and the reactions they create in an audience, Bamberg continued. Rather than studying text arrangement, for example, researchers might observe a speaker's body language. Word arrangement is still relevant, Bamberg noted, but it plays more of a supporting role, and the narrative is analyzed as a kind of performance. He added that the delivery of a message is also different in narratives expressed verbally and visually, so that, for example, a speaker wishing to induce feelings of empathy must behave in a particular way, while a speaker seeking forgiveness must look down to demonstrate regret. Thus, he asserted, the body and mind must work cooperatively to portray a convincing narrative.

Bamberg then identified three dimensions of narrative that are connected to the subject of power, each of which relates to value. First is agency, or the degree of control the character or narrator has in the situation. Bamberg explained that a character with low agency is one who is being affected primarily by external forces rather than exerting force on them, citing the example of someone who has somehow been violated. High and low agency are connected to such values as blame and responsibility, he noted, and there is a certain degree of moral order at stake when agency is navigated.

The second dimension Bamberg highlighted is the similarities and differences among various groups and individuals, studied in the disciplines of social linguistics and social psychology. He explained that this dimension encompasses group behavior, affiliations, belonging, and such categories as gender, race, and ethnicity.

Bamberg characterized the third dimension, temporal contour, as probably the most important, explaining that it includes states of change, development, and constancy. For example, he said, a character may express that he or she has—or has not—changed or grown in some way as the result of something that has happened. Temporal contour, he asserted, is the most relevant dimension in narrative because it can be found in all types of discourse.

Bamberg closed by briefly comparing master narratives and counternarratives. He explained that, like the big stories he had described in the first panel (see Chapter 2), master narratives are the background narratives used to make sense of the larger world. Without them, he added, people would not be able to communicate or make sense of the world. Counternarratives, on the other hand, are the narratives people create in their minds within the larger master narratives. Bamberg suggested that, while he is not a quantitative researcher, it would be worthwhile to study the relationship between master narratives and counternarratives using quantitative methods.

## DISCUSSION

Questions raised during the discussion focused on countering and inoculating the public against toxic narratives. A participant asked Louison Lavoy to discuss how the idea of inoculating the public from toxic narratives might be approached from the perspective of national security. One method that could be useful, she responded, is message testing. When designing a communications plan, she continued, marketers will test messages by buying ads that can be distributed easily and quickly online. If the message is well received online, she said, they will encourage influencers, customers, and employees to spread the message across their personal and professional networks. However, she noted, this is also one of the methods used by state and nonstate actors to spread false messages.

Louison Lavoy identified as another effective method what is called "chaining." Like message testing, she observed, chaining has also been used successfully to spread false messages. She cited the example of Russia, which used media chaining as way to disrupt the 2016 presidential election in a process that involves picking up messages from fringe groups and amplifying them just enough so that they are then picked up by other groups. The process is then repeated until the message moves from the fringe groups into more established groups, ultimately reaching mainstream society. Louison Lavoy added that individuals can also be brought into the chaining process if their actions are rewarded with additional social media attention (e.g., "likes," sharing content). However, she pointed out that if marketing techniques can be used to spread false messages, they can also be used to block and counter such messages. Johnson added that people with high credibility and social capital are also key to blocking and countering false messages. He has, for example, observed that when well-respected members of a group question the validity of a story, other members, even those with extreme views, will often do the same.

On the subject of inoculation, Cobb cited a recent study in which Israeli adolescents from conservative and liberal communities were exposed to stories that opposed the viewpoints with which they had grown up.[1] The researchers found, she said, that while the students from liberal communities changed their views after listening to the stories, the conservative students did not. She added that the liberal students also reported feeling better about themselves after hearing the conservative message. She highlighted as well the Seeds of Peace Summer Camp Program,[2] which brings Israeli and Palestinian adolescents together to encourage peace. She noted that before returning home, the adolescents must practice countering negative messages. With the use of scenario building and role-playing techniques, she explained, students can develop new narrative pathways that help preserve the understanding and empathy created by their experiences at camp. Johnson suggested that the way a story is told can make a difference in how it is received. Stories expressed using clear and easy-to-understand language, he observed, are more likely to be accepted by large audiences, adding that if the stories follow a clear narrative arc and are repeated continuously, they become even more compelling.

Raising the subject of behavioral economics, one participant asked the panelists what they thought of the use of nudge theory,[3] which suggests that a person's choices can be influenced by how information is presented, as a way of exerting power. Phelan responded that nudge is only one strategy available, and it is important to consider which strategy will work based on a particular situation. While the nudge strategy may be useful in some circumstances, he suggested, it may not be the right strategy in other cases. Louison Lavoy commented that, while she is not familiar with nudge theory, the term reminded her of the chaining method she had discussed earlier. Chaining, she elaborated, is a way of distributing propaganda so that it slowly creates a narrative pathway, and the narrative eventually becomes part of the audience's normal mental processing. Nudge theory reminded Bamberg of a series of social campaigns in China addressing such issues as aging. Although the videos used in this campaign are only about 60 seconds long and contain only music and visuals, he explained, they still manage to communicate very emotional and compelling narratives that are capable of influencing people.

Another participant brought up the clash between what she referred to as the scientific myth (i.e., scientific narratives) and other myths (e.g.,

---

[1] Porat, D.A. (2004). It's not written here, but this is what happened: Students' cultural comprehension of textbook narratives on the Israeli-Arab conflict. *American Educational Research Journal, 41*(4), 963–996.

[2] For more information on the Seeds of Peace Summer Camp Program, see https://www.seedsofpeace.org [April 2018].

[3] Thaler, R.H., Sunstein, C.R., and Pratt, S. (2014). *Nudge: Improving Decisions About Health, Wealth, and Happiness.* New York: Penguin Group.

more emotional narratives, such as those that relate to religion, politics, and culture). She suggested that someone standing in the scientific myth seeking to communicate effectively with someone standing in the religious myth must also stand in the religious myth. In other words, she argued, "if we are going to make any headway in this issue of power and narratives . . . we have to be able to stand within someone else's master narrative to be heard." Louison Lavoy suggested that one of the ways this might be done is for people to identify areas of commonality, such as shared values, and then communicate with each other based on those commonalities.

Another participant noted that some people have been so inundated with toxic narratives online that they no longer believe sources of information they once trusted. As a result, she argued, they are more susceptible to false messages that lead them to mistrust the Intelligence Community and other government agencies. She wondered how it might be possible to counter those messages. In response, Louison Lavoy suggested that, to counter and inoculate against false messages, society must find a way to rebuild a shared reality that will encourage positive forms of communication. Johnson added that it might be worthwhile to study available metadata to identify those people likely to be susceptible to these types of toxic narratives so that they can be inoculated against them. Louison Lavoy highlighted Ocean Protocol[4] as a recently developed framework holding data that could be used to measure susceptibility. However, she added, that information could be used for either positive or negative purposes.

---

[4] For more information on Ocean Protocol, see https://oceanprotocol.com [April 2018].

# 6

# Summative Comments

The workshop planning committee asked two members of the Intelligence Community (IC) to provide closing remarks. Both speakers have spent their entire career working with the IC and have extensive experience with intelligence analysis, moderator Carmen Medina, MedinAnalytics, explained. Following their remarks, the floor was opened for questions and discussion.

### REMARKS FROM KAREN MONAGHAN

Karen Monaghan, Central Intelligence Agency (retired), explained that, although she has never focused directly on the study of narrative, she has always considered understanding cultural narratives to be a crucial component of intelligence analysis. To understand a country and its leaders, she elaborated, an analyst must immerse himself or herself in that country's culture, adding that as a manager, she encouraged her analysts to listen to a country's music and read its literature to gain insight into its culture.

The workshop discussions spurred Monaghan's thinking about historical narratives and their important role in intelligence analysis. For example, she observed, historical narrative is key to knowing when and why a particular leader might change policies or take some other action, and can also shed light on how citizens might react to changes in government policies or to toxic narratives that are introduced into mainstream society. To illustrate this point, she noted that while many anthropologists were not surprised by the Rwandan genocide of 1994, many in the IC were. It was easier for anthropologists to anticipate the violent event, she explained, because they were familiar with Rwanda's historical narrative.

Monaghan is also interested in learning how to identify when a narrative is an indicator of a change in policy. She cited the example of Secretary of the Treasury Steve Mnuchin, who had made the statement earlier that day that a weaker dollar is good for the U.S. trade balance. However, she continued, prior administrations have always advocated for a strong dollar policy. Thus, she said, knowing whether Mnuchin's statement was just an off-the-cuff comment or part of a larger narrative signaling an official change in policy would be useful information for an analyst.

Referring to Mark Turner's comment that he is more interested in understanding the mental process of narratives than their content, Monaghan wondered whether the IC might be missing something when it focuses its attention on how specific words are used in messages. She noted that when examining Jihadi messages, for example, analysts try to predict future acts of violence by focusing on the terms used and what they might mean. She suggested that analysts might learn more if they thought more broadly about narrative.

At the same time, however, Monaghan raised the question of whether researchers collect data on what might be less obvious indicators of societal dissent, such as humor and jokes. Often, she suggested, people will express their thoughts and beliefs through humor rather than definitive statements, and collecting data on this type of communication could help identify sentiments that may otherwise be unapparent.

Monaghan identified as another area worth studying the universality of narratives, addressing such questions as "Do [Jihadi narratives] have universal appeal? How much are they being edited in order to appeal to different groups of Muslims in different countries?" She questioned whether there is a way to identify why some groups are more receptive to particular narratives and what might motivate a group to take or not take action. Answers to these types of questions, she argued, could be helpful in countering violent extremism.

In closing, Monaghan suggested that narrative research could also be an asset to analysts trying to supplement the predictions available from political polling outcomes, pointing to cases in which polling prior to an event did not forecast the actual outcome, such as Brexit or the 2016 U.S. presidential election. In particular, she suggested that studying narrative might provide insight as to why a particular narrative resonates with people and what effect it may have on how events unfold.

### REMARKS FROM JOSH KERBEL

As a former analyst and analytical methodologist, Josh Kerbel, National Intelligence University, said he appreciated learning more about narrative research and how it could benefit the IC. When people think about intelligence analysis, he noted, they often think about the parsing of collected data in search of answers. In his view, however, a more important aspect of analytical work—especially in a highly complex environment—is sensemaking (i.e., the creative development of holistic perspectives). To make sense of the stories that are held within the data collected, he elaborated, analysts must ask the right questions. He suggested that as a sensemaking tool, studying and understanding narratives could help analysts ask better questions, which in turn could help them provide more insightful findings to their clients (e.g., members of Congress, military officials, the President).

Although narrative is a part of analysis, Kerbel continued, it is not a term that analysts commonly use in relation to their own thinking. Instead, he said, they generally use the term "analytic lines," which refers to stories analysts tell each other and themselves to make sense of the areas they study. When information is received, he explained, it is usually measured against the prevailing analytic line to determine whether it is useful. To most analysts, in contrast, narratives are the stories told by adversaries, such as Al-Qaeda or ISIS.

Kerbel observed that studying narrative appears to be a more synthetic approach to analysis than the prevailing analytic approach typically used in the IC, in which the focus is on breaking down complex problems into more workable components. The synthetic approach, he explained, makes sense of information by pulling it together and shaping it into some kind of order, so it has the potential to be a very powerful tool for dealing with complexity. Thus, he argued, although it is not usually an explicit part of current analytic tradecraft and methodology, it should be, adding that the analytic culture and mindset may create a barrier to making that change. Narrative is inherently seen as not objective, he noted, because it is the expression of a story about something that happened, told from a certain perspective, yet objectivity, at least in theory, is central to the work of an analyst. Furthermore, he added, while narratives can be very rich and diffuse, analysts tend tomore write in a very clear, direct, and linear way. He noted as well that analytical work is guided by a set of core principles known as the Analytic Tradecraft Standards,[1] which require analysts to bring objectivity to their work and to express findings in a clear and logical way.

Kerbel mentioned two areas in which narrative could be particularly useful. The first is imaginative visualization. While the IC is accustomed to using visualization on the "back end" of analysis in its products and presentations, Kerbel explained, narrative may help them better visualize their work on the "front end" and thereby be a useful cognitive tool for making

---

[1] For more information on the Analytic Tradecraft Standards, see https://www.dni.gov/files/documents/ICD/ICD%20203%20Analytic%20Standards.pdf [March 2018].

sense of information. Second, he suggested, narrative could be helpful in making sense of the open-source social media or transmedia data discussed in earlier panels.

In closing, Kerbel suggested that narrative can be especially useful to analysts because it introduces a more human element into their thinking. He noted that analysts tend to observe humans from a distance, as if they were a "more physical kind of particles almost." If analysts could use narrative effectively as an analytic tool, he argued, it might help them think in more human terms, which might in turn improve the way they observe and understand others. The addition of this human element, he suggested, could then lead to more insightful conclusions.

## DISCUSSION

Medina opened the discussion by asking the panelists how they approach policy makers who enter an intelligence briefing with a set narrative that conflicts with the one they are about to present. Kerbel responded with a reminder that analysts often do not enter a briefing with a narrative in place; while they usually come equipped with collected data and facts, those facts are often not tied to a convincing narrative. However, he suggested, based on earlier panel discussions indicating that the best counter to a narrative is another narrative, analysts could be more persuasive if they prepared a narrative in advance of a briefing. Monaghan agreed, adding that analysts should know how their narratives compare with those of policy makers and where they may agree or differ. She suggested that parsing policy makers' narratives for them could help to identify holes in the narratives and open the door for analysts to introduce their findings.

One participant added that the National Geospatial-Intelligence Agency (NGA) has started to use narrative as a tool for delivering imagery analysis products in a more dynamic way. Kerbel agreed that some segments of the IC, such as NGA, are doing valuable research in this area, but observed that the findings of this research are not yet being applied in the day-to-day work of analysts across the IC, particularly in all-source agencies. Monaghan added that the Analytic Tradecraft Standards were written in a way that is somewhat antithetical to narrative.

One participant agreed with the earlier point that, instead of thinking of narrative as being opposed to objectivity, it might be helpful for analysts to think in terms of competing narratives: one may be better than another, but they both can exist as possible outcomes for the question at hand. Thus, the participant explained, the analyst is not giving up on the idea of objectivity, just shifting the view a bit. Kerbel agreed and responded that the IC is exploring the idea of competing narratives. However, he added, there is still a great deal of pressure to simplify the message. He noted that the role of an analyst is debated in the IC. Some believe, he observed, that it is an analyst's job to predict, stating that people who hold this view often reject the idea of a competing narrative because not only is their narrative based on what they believe will happen, but they feel competing narratives would not help clients who are seeking greater certainty from the IC. The other view, Kerbel continued, is that an analyst's job is not so much to predict but to inform the client of what is possible so the client can make a more informed decision. Those like himself who hold this latter view, he said, would likely agree that presenting competing narratives is an important part of an analyst's job.

Kerbel went on to point out that, regardless of differing views of competing narratives, analysts face pressure to present information in a short and simple format. A participant cautioned that to suggest there are no competing narratives could be dangerous because it implies there is only one correct narrative. Kerbel agreed and suggested that because analysts are increasingly dealing with complex phenomena, there is always more than one narrative for any given situation.

One participant suggested that topic modeling[2] might be helpful to the IC. Topic modeling, he explained, is an effective way of examining large amounts of data and extracting popular themes that are beginning to develop within a culture. He suggested that by using this technology to sort through the vast amounts of available open-source data, the IC might be able to identify evolving themes in subpopulations that might otherwise be overlooked. Another participant added that she recently had used a multilingual topic model to examine the French, English, and Kinyarwanda narratives from the Rwandan genocide. Running these types of computational linguistic models, she explained, can help researchers see the bigger picture that exists within the data, and qualitative methods can then be used to further clarify those findings. She added that the broad set of worldviews that emerges from the combination of these methods can then be used to develop theory from the ground up.

Kerbel seconded these comments and acknowledged that, because analysts frequently come from the more qualitative, social science disciplines, they often shy away from quantitative research. However, he added, the discussions that had taken place during the workshop might open the door for analysts' use of more quantitative approaches in their analyses. Medina suggested that one area of analysis in which quantitative methods might be particularly helpful is identifying triggers for violence. Currently, she noted, analysts tend to rely on their sources to inform them about events that are about to unfold. Even if a communication is intercepted, she said,

---

[2]For more information on topic modeling, see http://journalofdigitalhumanities.org/2-1/topic-modeling-a-basic-introduction-by-megan-r-brett [April 2018].

it must be a declarative statement for it to be recognized as identifying an upcoming event. Monaghan added that at times in her career, the information being collected about a country has appeared to suggest that a major change has occurred in the country's overall identifying narrative. However, she noted, changing the organizational view of that narrative is not easy; it does not get switched off like a light switch. Perhaps, she commented, with these methods there is a way to identify triggers that point to a shift in the narrative.

Another participant suggested that it would be helpful for researchers to have access to members of the IC so they could help shape the research questions that are addressed. He asked whether the panelists ever had opportunities to engage with researchers in this way. Monaghan and Kerbel agreed that this idea has merit, but noted both time and bureaucratic constraints, as well as concerns about the potential release of classified information, as reasons why analysts do not often collaborate with outside researchers. In the world of intelligence analysis, they added, the rewards are greater for those who focus on analytical tradecraft rather than time-consuming research. Kerbel said he believes the IC has only recently become more introspective about the process of analytical work, but this has not traditionally been an important focus for intelligence analysts.

One participant noted that he has had experience with trying to train members of the military in how to collect data and knows that people can find it difficult to adapt to new methods. He asked the panelists whether it would be helpful if members of the IC were trained in new techniques for working with narrative. Kerbel pointed out that obtaining buy-in from the IC on a new methodology would require understanding the IC culture, to which analysts have a strong connection. Monaghan suggested that a good way to introduce a new methodology that is tied to the IC culture is to obtain buy-in from leadership and identify someone willing to champion the new approach. She added that a new methodology is more likely to be accepted if it is applied to an area that is not related to secrets. For example, she suggested, analysts are more likely to need help finding ways to make sense of open-source data or to study a closed country, such as Cuba, where they lack personal sources of information. Another participant pointed out that using the term "mental models" instead of "narrative" might improve receptivity. Kerbel agreed and pointed out that many analysts may well be resistant to anything tied to the term "narrative." They work with narratives, he explained, but do not necessarily call them by that name. For example, he noted, "analytic lines" are narratives. He suggested that knowing how to phrase ideas is important and that this may be one of the most important lessons from the discussion.

# Appendix A

# Statement of Task for the Decadal Survey of Social and Behavioral Sciences for Applications to National Security

The National Academies of Sciences, Engineering, and Medicine will carry out a decadal survey on the social and behavioral sciences (SBS) in areas relevant to national security in two integrated phases. The first phase, a national summit (workshop), was completed in fall 2016. The statement of task for the second phase, a consensus process, is below.

An ad hoc consensus committee, drawing on membership from the summit steering committee, will be appointed to conduct the decadal survey aimed at identifying opportunities that are poised to contribute significantly to the Intelligence Community's (IC's) analytic responsibilities. The study will identify opportunities throughout the social sciences (e.g., sociology, demography, political science, economics, and anthropology) and from behavioral sciences (e.g., psychology, cognition, and neuroscience) and will draw on discussions at the summit to frame its inquiry. Attention will also be paid to work in allied professional disciplines, such as engineering, business, and law, and a full variety of cross-disciplinary, historical, case study, participant, and phronetic approaches.

The committee will work with the Office of the Director of National Intelligence and security community members to understand government needs and expectations. The final report will be based on the committee's consideration of broad national security priorities; relevant capabilities of elements within the security community to support and apply SBS research findings; cost and technical readiness; likely growth of research programs; emerging SBS data, procedures, personnel, and other resources; and opportunities to leverage related research activities not directly supported by government. The committee will specify a range of relevant work that could be useful to the IC for their consideration in developing future research priorities.

The committee's primary tasks will be to:

1. Assess progress in addressing selected major social and behavioral scientific challenges that might prove useful to national security. Include discussion of approaches that are gaining strength and those that are losing strength.
2. Identify SBS opportunities that can be used to guide security community investment decisions and application efforts over the next 10 years.
3. Specify approaches to facilitate productive interchange between the security community and the external social science research community.
4. Reflect on the application of the decadal model to the SBSs and identify lessons learned (insights into how to approach and perform the decadal survey process) and promising practices (activities that could facilitate future decadal surveys in the SBSs and similar disciplines and maximize their ultimate utilities to sponsors and the scientific community).

# Appendix B

# Workshop Agenda

UNDERSTANDING NARRATIVES FOR NATIONAL SECURITY
PURPOSES: A WORKSHOP
January 24, 2018

Keck Center
500 Fifth Street, NW
Washington, DC
Room 101

| | |
|---|---|
| 8:30 a.m. | **Workshop Registration Opens** |
| 9:00 a.m. | **Workshops Commence** |
| 9:00 a.m. | **Welcome and Overview of Events**<br>*Sujeeta Bhatt*, Study Director<br>    Audience information<br>*Paul Sackett*, University of Minnesota, SBS Decadal Survey Chair<br>    Welcome<br>*William "Bruno" Millonig*, Acting Director of National Intelligence for Science and Technology, Office of the Director of National Intelligence<br>    Sponsor perspective and context for study and workshops |
| 9:30 a.m. | **Opening Remarks**<br>*Carmen Medina*, MedinAnalytics, LLC, Workshop Committee Chair |

### Research Panel Presentations and Discussion

**9:35 a.m.** **Panel 1: Introduction to Narrative Research in the Social and Behavioral Sciences**
Moderators: *Jeffrey Johnson*, University of Florida, and *Betty Sue Flowers*, University of Texas, Austin

Speakers will present on the state-of-the-art in narrative studies and examine cutting-edge questions relevant to national security and intelligence analysis.

*Roberto Franzosi*, Emory University
*Mark Turner*, Case Western Reserve University
*James Pennebaker*, University of Texas, Austin
*Michael Bamberg*, Clark University

**10:30 a.m.** **Discussion and Q&A**

**10:45 a.m.** **BREAK**

**10:55 a.m.** **Panel 2: Why study narrative? What is the comparative advantage of the study of narrative for national security? How might we study/track narratives comprehensively in real time?**
Moderators: *Carmen Medina*, MedinAnalytics, LLC, and *David Matsumoto*, San Francisco State University

*Michael Dahlstrom*, Iowa State University
*Pauline Cheong*, Arizona State University
*Mark Turner*, Case Western Reserve University

**11:50 a.m.** **Discussion and Q&A**

**12:05 p.m.** **LUNCH**

**1:05 p.m.** **Panel 3: How might artificial intelligence and other emerging technologies affect narratives and their formation? What is the impact of social media on the development and dissemination of narratives?**
Moderators: *Sara Cobb*, George Mason University, and *Doug Randall*, Protagonist

*Michael Young*, University of Utah
*Catherine Tejeda*, Parenthetic

| | |
|---|---|
| 2:00 p.m. | **Discussion and Q&A** |
| 2:15 p.m. | **BREAK** |
| 2:25 p.m. | **Panel 4: What is the relation of narrative to power? How does narrative work as a tool for mobilization and intervention? How do different societies' narratives vary and sometimes clash?**<br>Moderators: *Jeffrey Johnson*, University of Florida, and *Sara Cobb*, George Mason University<br><br>*James Phelan*, Ohio State University<br>*Debra Louison Lavoy*, Narrative Builders<br>*Michael Bamberg*, Clark University |
| 3:20 p.m. | **Summative Comments**<br>*Karen Monaghan*, Central Intelligence Agency (retired)<br>*Josh Kerbel*, National Intelligence University |
| 3:50 p.m. | **Discussion and Q&A** |
| 4:50 p.m. | **Closing Remarks**<br>*Carmen Medina*, MedinAnalytics, LLC, Workshop Committee Chair |
| 5:00 p.m. | **ADJOURN** |

# Appendix C

# Participants List

This list includes participants across all three workshops held January 24, 2018, to gather information for the Decadal Survey of Social and Behavioral Sciences for Applications to National Security. (*presenters)

Vincent Alcazar
U.S. Air Force

Adam Amos-Binks
North Carolina State University

*Edward Awh
University of Chicago

*Michael Bamberg
Clark University

James Belanich
Institute for Defense Analyses

Gina Bennett
U.S. Government

Elizabeth Bernstein
National Geospatial Intelligence Agency

Sujeeta Bhatt
National Academies

Laura Biggerstaff
Office of Naval Research

Kathleen Carley
Carnegie Mellon University

*Remco Chang
Tufts University

*Pauline Cheong
Arizona State University

*Ted Clark
CENTRA Technology, Inc.

Noshir Contractor
Northwestern University

Chris Cox
Defense Intelligence Agency

Thelma Cox
National Academies

Bruce Crawford
Independent Researcher

*Michael Dahlstrom
Iowa State University

Barbara Anne Dosher
University of California, Irvine

*David Dunning
University of Michigan

Christine Dye
C2 Technologies, Inc.

Eric Eisenberg
University of South Florida

*Jill Ellingson
University of Kansas

*Thomas Fingar
Stanford University

*Steve Fiore
University of Central Florida

*Roberto Franzosi
Emory University

Paul Gade
George Washington University

Angela Gonsorcik
Federal Bureau of Investigation/
  National Intelligence University

Joseph Gordon
National Intelligence University

Justin Grossman
U.S. Department of Defense

Thomas Guarrieri
University of Maryland

David Gwinn
U.S. Navy

Craig Haimson
U.S. Department of Defense

J.C. Herz
Ion Channel, Inc.

David Honey
Defense Advanced Research
  Projects Agency

John Hoven
Independent

Alexis Jeannotte
Intelligence Advanced Research
  Projects Agency

Jeffrey C. Johnson
University of Florida

Sallie Keller
Virginia Tech

APPENDIX C

*Josh Kerbel
National Intelligence University

Ben King
C2 Technologies, Inc.

*Debra Louison Lavoy
Narrative Builders

John Mathews
U.S. Department of Defense

David Matsumoto
San Francisco State University

*Roger Mayer
North Carolina State University

Shana McLean
The MITRE Corporation

Carmen Medina
MedinAnalytics, LLC

*Barbara Mellers
University of Pennsylvania

Carmen Meyer
U.S. Department of Defense

*Karen Monaghan
Central Intelligence Agency
    (retired)

Roy Monfort
JIDO/Leidos

Fran P. Moore
CENTRA Technology, Inc.

Kent Myers
Office of the Director of
    National Intelligence

Brian Nordmann
U.S. Department of State

Kevin O'Connell
Innovative Analytics and
    Training, LLC

William O'Hara
U.S. Department of Defense

Mark Orr
Virginia Tech

*Scott E. Page
University of Michigan

Henry Peltokangas
Cisco

*James Pennebaker
University of Texas, Austin

*James Phelan
Ohio State University

*Peter L.T. Pirolli
Institute for Human &
    Machine Cognition

Emma Price
Leidos

Joy Rohde
University of Michigan

Scott Ruston
Arizona State University

Paul R. Sackett
University of Minnesota

Laura Sappelsa
ANSER

Monique Sargeant
University of Maryland
   University College

Michael Schrage
Massachusetts Institute
   of Technology

Julie Schuck
National Academies

Ben Shneiderman
University of Maryland

Eric Sid
National Institutes of Health

Erin Sikorsky
Director of National Intelligence/
   Strategic Futures Group

Stacey Smit
National Academies

Laura Steckman
The MITRE Corporation

Timothy Stephens
Ballpark

*Victoria Stodden
University of Illinois

*Danielle Albers Szafir
University of Colorado Boulder

Jeffrey Taliaferro
Tufts University

*Catherine Tejeda
Parenthetic

Jorhena Thomas
Georgetown University

*Nancy Tippins
The Nancy T. Tippins Group, LLC

Elizabeth Townsend
National Academies

*Mark Turner
Case Western Reserve University

Garrett Tyson
National Academies

Marina Ulmishek
U.S. Government

*Brian Uzzi
Northwestern University

Drew Vandeth
IBM Research

Kate Von Holle
University of Chicago

Barbara Wanchisen
National Academies

*Adam Waytz
Northwestern University

*Alyson Wilson
North Carolina State University

Renée L. Wilson Gaines
National Academies

Leah Windsor
University of Memphis

Jeremy Wolfe
Brigham and Women's Hospital,
 Harvard Medical School

Colin Wood
Engineer Research and
 Development Center/U.S. Army
 Corps of Engineers

*Michael Young
University of Utah

*Andrew Ysursa
Salesforce, Inc.

*Steve Zaccaro
George Mason University

Mark Zimmermann
The MITRE Corporation

# Appendix D

# Biographical Sketches of Steering Committee Members and Presenters

**Michael Bamberg** (*Presenter*) is professor of psychology, cultural studies and communication, and English at Clark University. He has contributed varied strands to developmental psychology, applied linguistics, and identity theory. At Clark, he has been instrumental in advancing undergraduate and graduate training in qualitative research. He also has been heavily involved in the establishment of the Society for Qualitative Inquiry in Psychology within the American Psychological Association. He is nationally and internationally known for teaching workshops on qualitative methods and narrative analysis and recently held an appointment as Yunshan Chair professor at the Guangdong University of Foreign Studies. As a developmentalist, Dr. Bamberg investigates identity formation processes as processes situated in context and in interaction. He is particularly interested in the question of student and teacher identity, in the role of narrative in organizational/institutional identity formation processes, and in the concept of "branding." He received a Ph.D. in psychology from the University of California, Berkeley.

**Sujeeta Bhatt** (*Study Director*) is a senior program officer with the National Academies of Sciences, Engineering, and Medicine and study director for the Decadal Survey of Social and Behavioral Sciences for Applications to National Security. She was formerly a research scientist at the Defense Intelligence Agency (DIA) and was detailed to the Federal Bureau of Investigation's High-Value Detainee Interrogation Group (HIG). Prior to that, she was an assistant professor in the Department of Radiology at the Georgetown University Medical Center on detail to DIA/HIG. Her work at DIA and HIG entailed identifying knowledge gaps and developing and managing research projects to address those gaps. Her work in the Intelligence Community focused on the psychological and neuroscience bases for credibility assessment, biometrics, insider threat, intelligence interviewing and interrogation methods, and the development of research-to-practice modules on interrogation-related topics to promote the use of evidence-based practice in interviews/interrogations. She holds a Ph.D. in behavioral neuroscience from American University.

**Pauline Cheong** (*Presenter*) is professor at the Hugh Downs School of Human Communication, Arizona State University (ASU). She studies the complex interactions between communication technologies and different cultural communities around the world. Her recent grant-funded projects related to changing knowledge, authority, and leadership practices have examined how clergy and teachers maintain the interest of their students and congregations when the use of mobile and social media is so prevalent. Another of her interests is how religious groups use technology to interact and form local and global communities. She has investigated how communication technologies facilitate and constrain relations within cyber-vigilante groups and rumor-mongers in contested narrative landscapes, as well as how underserved and youth populations experience multiple digital divides. Dr. Cheong has published more than 80 articles and books and has received research awards from the National Communication Association, Western Communication Association, and International Communication Association. She serves on national and international boards and committees and has chaired doctoral colloquiums. At ASU's Center for Asian Research, she is co-director of @AsiaMediated, funded by the U.S. Department of Education. Dr. Cheong received an M.A. and a Ph.D. from the University of Southern California.

**Sara Cobb** (*Steering Committee*) is a professor at the School for Conflict Analysis and Resolution (S-CAR) at George Mason University, where she was also the director for 8 years. She teaches and conducts research on the relationship between narrative and violent conflict. She is also director of the Center for the Study of Narrative and Conflict Resolution at S-CAR, which provides a hub for scholarship on narrative approaches to conflict analysis and resolution. Formerly, she was director of the Program on Negotiation at Harvard Law School, and she has held positions at a variety of tier-one research institutions. She has also consulted to and/or conducted training for a host of public and private organizations. Dr. Cobb is widely published. She has been a leader in the fields of negotiation and conflict resolution, conducting research on the practice of neutrality, as well as the production of "turning points" and "critical moments" in negotiation processes. Some of this research is based on case studies from her field research in Guatemala, Chile, Rwanda, and the Netherlands. The blend of academic research, program development, and practice enables her to develop research projects that can yield practical understanding and generate effective interventions. She received a Ph.D. from the University of Massachusetts Amherst.

**Michael Dahlstrom** is associate director of the Greenlee School of Journalism and Communication and an associate professor at Iowa State University. His research focuses on the effects of narratives on perceptions of science. He is co-editor of an upcoming edited volume focusing on the often overlooked ethical challenges underlying science communication. He is also a past head of the Communicating Science, Health, Environment and Risk Division of the Association for Education in Journalism and Mass Communication, and received the Shakeshaft Master Teaching Award in 2013. He earned an M.S. in biophysics from Iowa State University and a joint Ph.D. in journalism and mass communication and environmental resources from the University of Wisconsin–Madison.

**Betty Sue Flowers** (*Steering Committee*) is former director of the LBJ Presidential Library and Museum and former Kelleher professor of English and member of the Distinguished Teachers Academy at the University of Texas, Austin, as well as distinguished alumnus. She is also a poet, editor, and business consultant. She has served as a moderator for executive seminars at the Aspen Institute for Humanistic Studies; a consultant for the National Aeronautics and Space Administration; a visiting advisor to the secretary of the Navy; public director of the American Institute of Architects; and editor for Shell International in London, the Organization of American States, and the World Business Council in Geneva. Dr. Flowers received a B.A. and an M.A. from the University of Texas, Austin, and a Ph.D. from the University of London.

**Roberto Franzosi** (*Presenter*) is professor of sociology and linguistics at Emory University. His main substantive interest has been in social protest, with projects on Italian strikes, on the rise of Italian fascism (1919–1922), and on racial violence in Georgia (1875–1930). He has had a long-standing methodological interest in issues of language and measurement of meaning in texts. He is currently working on automatic computational linguistics approaches to the extraction of social actors and their actions from narrative texts (SVO, or Subject-Verb-Object) and automatic ways of visualizing the shape of stories. Dr. Franzosi received a B.A. in literature from the University of Genoa and a Ph.D. in sociology from Johns Hopkins University.

**Jeffrey Johnson** (*Steering Committee*) is a professor of anthropology at the University of Florida. He is also an adjunct professor in the Institute for Software Research at Carnegie Mellon University. He was director of the Summer Institute for Research Design in Cultural Anthropology from 1996 to 2015. He is a former program manager with the Army Research Office (IPA), where he started the basic science research program in the social sciences. He has conducted extensive long-term research comparing group dynamics and the evolution of social networks among overwintering crews at the American South Pole Station and at the Polish, Russian, Chinese, and Indian Antarctic stations. Using these isolated human group settings as space analogs, he is currently studying aspects of team cognition relative to mission success. He has served as editor, co-editor, or associate editor of several journals. He received his Ph.D. in social science from the University of California, Irvine.

**Josh Kerbel** (*Presenter*) is a member of the research faculty at the National Intelligence University (NIU). His primary research focus at NIU is the future of intelligence analysis, especially how the Intelligence Community (IC) can better anticipate the emergent dynamics spawned by an increasingly complex security environment. More specifically, he explores the disruptive innovations demanded of the IC by a security environment that is fundamentally different from the Cold War environment that still profoundly—and problematically—shapes its legacy mindsets, processes, and habits. Prior to joining NIU, Mr. Kerbel held senior analytically focused positions in the Defense Intelligence Agency, the Office of the Director of National Intelligence, the U.S. Navy, the Central Intelligence Agency, and the Office of Naval Intelligence. He holds degrees from the George Washington University and the London School of Economics, as well as professional certifications from the Naval War College and the Naval Postgraduate School. He was also a postgraduate (Seminar XXI) fellow in the Center for International Studies at the Massachusetts Institute of Technology.

**Debra Louison Lavoy** (*Presenter*) is a marketing executive who has been studying narrative explicitly since 2010. She holds a degree in computer science and neurobiology from McGill University and spent the first decade of her career as a software engineer in research and commercial settings. Her narrative research grew out of roles at complex technology companies that struggled to explain the capabilities of their technology and why it mattered. Two years ago, she founded a boutique consulting firm, Narrative Builders, whose clients include a wide range of B2B, B2C, and nongovernmental organizations that now have well-developed and well-deployed organizational narratives that win investment and market share and build exceptional teams.

APPENDIX D

**David Matsumoto** (*Steering Committee*) is a professor of social psychology at San Francisco State University and director of the Culture and Emotion Research Lab, which focuses on studies involving culture, emotion, social interaction, and communication. He is well known for his work in the field of microexpressions, facial expression, gesture, and nonverbal behavior. He has served as editor-in-chief, editor, or editorial board member for several journals. He holds a Ph.D. in psychology from the University of California, Berkeley.

**Carmen Medina** (*Steering Committee Chair*) is the founder of MedinAnalytics, LLC, which provides analytic services on national security issues, cognitive diversity, global trends, and intrapreneurship. From 2005 to 2007, she was part of the executive team that led the Central Intelligence Agency's (CIA's) Analysis Directorate. In her last assignment before retiring, she oversaw the CIA's Lessons Learned program and led the agency's first effort to address the challenges posed by social networks, digital ubiquity, and the emerging culture of collaboration. She was a leader on diversity issues at the CIA, serving on equity boards at all organizational levels and across directorates. She was the first CIA executive to conceptualize many information technology applications now used by analysts, including online production, collaborative tools, and Intellipedia. Upon her retirement from the CIA, she received the Distinguished Career Intelligence Medal. From 2011 to 2015, she was a member of Deloitte Federal Consulting, where she served as senior advisor and mentor to Deloitte's flagship innovation program, GovLab. She holds a B.A. in comparative government from the Catholic University of America.

**William "Bruno" Millonig** (*Presenter*) is Acting Director of National Intelligence for Science and Technology in the Office of the Assistant Director for Acquisition, Technology and Facilities at the Office of the Director of National Intelligence. Appointed in November 2017, Mr. Millonig is responsible for guiding the Intelligence Community's (IC's) scientific and technological integration through effective strategies, policies, and programs that ultimately allow the IC to close intelligence gaps. Prior to this position, he oversaw the Defense Intelligence Agency headquarters' research and development, technical collection, and analytic responsibilities in support of the nation's space and counterspace situational awareness. He also served as chief, National Measurement and Signature Intelligence (MASINT) Office, and chairman, National MASINT Committee. A command pilot with more than 4,800 flight hours, he retired from the U.S. Air Force in 2009 as director of strategic planning for homeland defense and counterterrorism issues. He is a Distinguished Flying Cross recipient, was commander of the U.S. Air Force's training squadron of the year (2004), and holds numerous team and individual awards. He graduated from the U.S. Air Force Academy with a B.S. in engineering and earned master's degrees in aviation operations and management from Embry Riddle University and in strategic studies from the U.S. Army War College.

**Karen Monaghan** (*Presenter*) is a former senior federal executive with the Intelligence Community. She retired in 2017 after 32 years of service, largely at the Central Intelligence Agency. She is known for being a strategic thinker and writer with broad multidisciplinary and geographic knowledge and experience analyzing economic and political developments in Asia, Africa, Europe, and Latin America.

**James Pennebaker** (*Presenter*) is Regents professor of psychology and executive director of a university-wide educational initiative called Project 2021 at the University of Texas, Austin. His earliest work examined the psychology of physical symptoms. That research ultimately led to his discovery of expressive writing—that writing about emotional upheavals improves physical health and immune function. More recent studies have explored natural language. Dr. Pennebaker's research has revealed that everyday word use is related to personality, deception, status, group dynamics, and emotional states. He is now working with his university's senior administration to rethink 21st-century education models. He has received several university and international awards for his research and teaching.

**James Phelan** (*Presenter*) is distinguished university professor of English at Ohio State University and the recipient of numerous prizes and grants, including an honorary doctorate from Aarhus University in 2013. He has also taught at the University of Toronto, Colorado College, and Norway's Centre for Advanced Study in Oslo. In 2013, the International Society for the Study of Narrative named its annual award for best essay in *Narrative* the James Phelan Prize. Dr. Phelan is one of the founding members of Project Narrative at Ohio State, a unit recognized internationally as the premier center for research and teaching in narrative theory. His writings develop and deploy a rhetorical theory of narrative, one rooted in the principle that storytelling is an action in which one or more tellers recount events for one or more audiences to fulfill some particular purposes. He received an M.A. and a Ph.D. in English language and literature from the University of Chicago.

**Doug Randall** (*Steering Committee*) is founder and CEO of Protagonist, a high-growth narrative analytics company. Protagonist mines beliefs in order to energize brands, win narrative battles, and understand target audiences.

It uses natural language processing, machine learning, and deep human expertise to identify, measure, and shape narratives. The Protagonist platform was built on 10 years of narrative science that was initially developed to improve the American brand around the world for the U.S. government. Today, it is used by dozens of the world's leading chief marketing officers, business leaders, and foundations. Mr. Randall has lectured on a number of topics at the Wharton School, Stanford University, and the National Defense University. He was previously a partner at Monitor, founder of Monitor 360, and co-head of the consulting practice at Global Business Network. Before that, he was a vice president at Snapfish, a senior consultant at Decision Strategies, Inc., and a senior research fellow at the Wharton School. Mr. Randall received a B.A. from the University of Pennsylvania and an M.B.A. from the Wharton School.

**Paul Sackett** (*Decadal Survey Chair*) is the Beverly and Richard Fink distinguished professor of psychology and liberal arts at the University of Minnesota. His research interests revolve around various aspects of testing and assessment in workplace, educational, and military settings. He has served as president of the Society for Industrial and Organizational Psychology, as co-chair of the committee producing the Standards for Educational and Psychological Testing, as a member of the National Academies of Sciences, Engineering, and Medicine's Board on Testing and Assessment, as chair of the American Psychological Association's (APA's) Committee on Psychological Tests and Assessments, and as chair of APA's Board of Scientific Affairs. He holds a Ph.D. in industrial/organizational psychology from Ohio State University.

**Catherine Tejeda** (*Presenter*) is founder and CEO of Parenthetic, a small business that aims to bring more science to the art of influence. Parenthetic grew from her passion for developing techniques and technologies to tackle difficult challenges with global impact. She has more than a decade of experience studying and analyzing behavior across cultures, as well as methods for understanding and influencing attitudes and behaviors. She has worked at every stage of the communication development cycle: strategy and design; audience, competitor, and communication landscape analysis; message content creation, distribution, and assessment; and quantifying of campaign effects. She has worked across the public and private sectors on projects spanning counterproliferation, election monitoring, domestic and foreign extremism, health care and health crises, product and market forecasting, and luxury product sales. At Parenthetic, she leads teams for public- and private-sector clients focused on creating replicable processes for influencing individuals and populations measurably and more effectively. She applies expertise to advance the way people develop, discover, isolate, and measure the effect of influence campaigns.

**Mark Turner** (*Presenter*) is founding director of the Cognitive Science Network and co-director of the Red Hen Lab. He has received the Anneliese Maier Research Prize from the Alexander von Humboldt Foundation and the Prix du Rayonnement de la Langue et de la Litérature Françaises from the French Academy. He is founding president of the Myrifield Institute for Cognition and the Arts. He is a fellow of the Institute for Advanced Study, the Center for Advanced Study in the Behavioral Sciences, the National Humanities Center, the John Simon Guggenheim Memorial Foundation, the Institute of Advanced Study at Durham University, the Centre for Advanced Study at the Norwegian Academy of Science and Letters, the New England Institute for Cognitive Science and Evolutionary Psychology, the National Endowment for the Humanities, and the Institute for the Science of Origins. He is also an extraordinary member of the Humanwissenschaftliches Zentrum der Ludwig-Maximilians-Universitätm, an external research professor of the Krasnow Institute for Advanced Study, and a distinguished visiting professor at Hunan Normal University.

**Michael Young** (*Presenter*) is a professor in the School of Computing and deputy director of the Entertainment Arts and Engineering Program at the University of Utah in Salt Lake City, where he directs the Liquid Narrative research group. He works to develop computational models of interactive narrative with applications to computer games, educational and training systems, and intelligence analysis. His work is grounded in computational approaches, but he seeks to cross disciplinary boundaries, involving collaborators and concepts from cognitive psychology, linguistics, narrative theory, cinematography, and other disciplines in which human cognition and interaction are central. He is an Association for Computing Machinery distinguished scientist and a senior member of both the Association for the Advancement of Artificial Intelligence and the Institute of Electrical and Electronics Engineers. In 2017, he was elected to the inaugural class of Higher Education Video Game Alliance fellows.